T0342349

Unnatural Disasters

Unnatural Disasters

Why Most Responses to Risk and Climate Change Fail but Some Succeed

Gonzalo Lizarralde

Columbia University Press
New York

Columbia University Press
Publishers Since 1893
New York   Chichester, West Sussex
cup.columbia.edu

Library of Congress Cataloging-in-Publication Data
Names: Lizarralde, Gonzalo, 1974– author.
Title: Unnatural disasters : why most responses to risk and climate change fail but
some succeed / Gonzalo Lizarralde.
Description: New York : Columbia University Press, [2021] | Includes bibliographical
references and index.
Identifiers: LCCN 2020055429 (print) | LCCN 2020055430 (ebook) |
ISBN 9780231198103 (hardback) | ISBN 9780231552509 (ebook)
Subjects: LCSH: Environmental disasters. | Climatic changes. | Sustainability. | Green
movement. | Sustainable development.
Classification: LCC GE146 .L59 2021 (print) | LCC GE146 (ebook) |
DDC 363.34/1–dc23
LC record available at https://lccn.loc.gov/2020055429
LC ebook record available at https://lccn.loc.gov/2020055430

Columbia University Press books are printed
on permanent and durable acid-free paper.
Printed in the United States of America

Cover design: Elliott S. Cairns
Cover image: Simone Gilioli / stock.adobe.com

To my Natalia and my Sebastian

Contents

Preface

Telling stories of human suffering is never easy. How can I call someone a victim without victimizing her? Or highlight people's strength without patronizing them? How can I state, for example, that one hundred children died without reducing them—their unique lives, their parents' pain—to a statistic? In this book, I have tried to face this challenge in two ways.

One way is by applying rigorous science to the way my colleagues and I captured, selected, and eventually translated the stories I tell here. This is not, though, the type of academic book where you will find an exhaustive review of the theory of disaster studies. Don't expect to find a long section on research methods or a fancy new model of disaster prevention. Nor is this book a simple collection of anecdotes or an assortment of personal experiences. Instead, I explore a selected number of popular concepts against the background of real-life cases and human stories. Thus, the narratives you will read in this book are not casual snapshots of people's lives. They illustrate patterns that my colleagues and I have found after applying rigorous social science methods to more than forty cases of human settlements and territories.

The second way is by remembering that reading and writing about tragedies are also ways of empathizing with the people affected by them. In a world of frequent destruction, there are so many urgent things to say and do—surely it is important to write about them.

For the past twenty years, I have worked with people affected by disasters, war, and the effects of climate change. I have witnessed not only human suffering but also the work of hundreds of people who attempt to create better and safer living conditions for all. I have also observed the emergence of problems that motivated me to write this book.

For years, scientists have made pertinent contributions to our understanding of risks, disasters, and reconstruction. We must celebrate these ideas and theories. They have helped to bring the goals of sustainability, resilience, adaptation, and "building back better" to city halls and board meetings. But we must also acknowledge that as academic ideas have gained popularity, they have been taken to ever higher levels of abstraction. They have been crudely generalized and blindly adopted in increasingly diverse contexts and by very different people. Today, the result is a blunt disconnect of concepts from the very realities they attempt to explain.

While these academic concepts have been rapidly adopted in disaster studies and practice, they have become a cloak for hidden agendas. Increasing numbers of academics, consultants, politicians, professionals, and technocrats promote a package of concepts and ideas as virtuous actions, failing to address their limits, cascading effects, blind spots, and unintended consequences. Scholars, representatives of multinational organizations, and international consultants have transformed concepts into normative frameworks to be applied indiscriminately, masking the conditions that create winners and losers during implementation. They rely on them for judgment and rapid decision making, missing the opportunity to produce solutions tailored to the needs and expectations of individuals and specific social groups. Meanwhile, many useful (but less popular) ideas have been ignored.

In writing this book, I take the risk of being too academic for some and not scholarly enough for others. I take this risk knowing that there is an urgent need to close the gap between the abstract concepts adopted in disaster studies and real-life situations. I take it convinced that we must bring abstract concepts back to earth—back to where human stories belong.

Acknowledgments

This book is unfairly signed by one single author. In fact, most of the arguments presented here are the result of collective, not individual, research and thinking. I am particularly grateful to the following:

—Kate Stern, who provided important suggestions for the arguments exposed in this book and revised the whole thing.
—Robert Lecker, my literary agent, who guided me during the process.
—Jon Turney, who helped me identify the scope of the book and believed in it even before I did.
—Lisa Bornstein, Isabelle Thomas, Kevin Gould, Christopher Bryant, Danielle Labbé, Gabriel Fauveaud, David Wachsmuth, Benjamin Herazo, and all the members and students affiliated to the Observatoire universitaire de la vulnérabilité, la résilience et la reconstruction durable (Œuvre durable, or the Disaster Resilience and Sustainable Reconstruction Research Alliance) for their contributions to the arguments exposed here.

—All the researchers, universities, and students affiliated to the ADAPTO project (Climate Change Adaptation in Informal Settings: Understanding and Reinforcing Bottom-Up Initiatives in Latin America and the Caribbean), funded by IDRC Canada, for their contributions to the development of several arguments presented in this book, for their case studies, and for the empirical work conducted in Latin America and the Caribbean.

—Mauro Cossu, Georgia Cardosi, Faten Kikano, Mahmood Fayazi, Anne-Marie Petter, Lisa Hasan, Tapan Dhar, and all the students affiliated to the IF Research Group (grif) for their discussions, research results, and pertinent ideas for this book.

—All members and partners of the i-Rec (Information and Research for Reconstruction) community, for their inspiration and help in developing innovative ideas.

—All students and professors involved in our Œuvre durable/i-Rec online debates and student competitions, for their inspiring ideas and creativity.

—Daniel Pearl, David Ross, Jean-Jacques Terrin, Jean-Paul Boudreau, Thomas Schweitzer, Anne Marchand, Carlo Carbone, and all the members of the Fayolle-Magil Construction Research Chair in Architecture, the Built Environment and Sustainability, Université de Montréal, for their contributions to the studies on climate change mitigation.

—All the members of the EU-funded project in Port-au-Prince, including Jean Goulet, Paul Bodson, Paul Martel, Yona Jébrak, and all the colleagues at UQAM for their contributions to the case study in Haiti and arguments about reconstruction.

—All the professors I interviewed to clarify arguments and ideas.

—The students of the Masters in Architecture (Université de Montréal) and the professors and students involved in the international workshops we have conducted since 2012.

—The administrative staff, lecturers, and professors at: Université de Montréal, Canada; McGill University, Montréal, Canada; Concordia University, Montréal, Canada; Universidad del Valle, Cali, Colombia; Universidad del Bío-Bío, Concepción, Chile; Universidad Central

"Marta Abreu" las Villas, Santa Clara, Cuba; Universidad Javeriana, Bogotá, Colombia; Université d'État d'Haïti, Port-au-Prince, Haiti; FLACSO, Quito, Ecuador; and the NGO Corporación Antioquia Presente, Medellín, Colombia, for their support to empirical research and academic activities.

—The residents of informal settlements, local leaders, and officers of several public and private organizations that were interviewed during the development of the research projects that led to this publication.

—Ksenia Chmutina, Lee Bosher, Andrew Dainty, and other colleagues at Loughborough University, UK, for their support to our research on sustainability and resilience.

—Claude Paquin and the staff at FORMES, for their contribution to our INTREFACES seminars.

—All participants in our INTREFACES seminars, for their ideas and pertinent arguments.

—The members of the RIISQ network in Quebec (funded by FRQNT), for the visibility and networking for our research activities.

I acknowledge the contribution of the institutions that funded the research initiatives that permitted creating the case studies presented in this publication, notably:

—Canada's International Development Research Centre, for its funding of our case studies and implementation initiatives in Latin America and the Caribbean.

—The Social Sciences and Humanities Research Council—Canada, for funding the research work that led to the case studies in Colombia, Cuba, Haiti, Chile, Vietnam, the United States, and other places.

—The Fonds de recherche du Québec—société et culture FRQSC, for its contribution to Œuvre durable.

—The Université de Montréal (Direction des affaires internationales), for its support of international workshops and studios.

—The Fayolle-Magil Construction Research Chair in Architecture, the Built Environment and Sustainability, Université de Montréal, for the support of studies in Canada.

—The European Union, the Université d'État d'Haïti, and the American Red Cross for their support to our work in Haiti.
—The City of Montreal for its support to the study on quality in public projects in Canada.
—The Maison du développement durable, for its support to the study on green buildings in Canada.
—Loughborough University, UK, for its support to our research on sustainability and resilience.

Unnatural Disasters

Introduction

"It Won't Be Easy, But We Have Little Choice"

Is it not ridiculous for chance to dictate someone's death and have the circumstances of that death—secrecy, publicity, the fixed time of an hour or a century—not subject to chance?

. . . Because Babylon is nothing else than an infinite game of chance.

—Jorge Luis Borges, "The Lottery in Babylon"

THE SOUND OF BUILDINGS FALLING

When an earthquake destroyed the city of Port-au-Prince in 2010, Haitians heard—but mostly felt—a mighty noise. They called it *Goudougoudou*, a clever Creole onomatopoeia. There is no English word that can evoke the vibration of earth cracking, the clamor of people suffering, and the sound of buildings falling,[1] so I will simply refer to it as a disaster. But, as you can guess from the title of this book, I won't call it a *natural* disaster.

Months later, in June of that same year, many Port-au-Princiens still sensed the Goudougoudou in their heads. Rubble lined the streets. Electricity was unreliable and drinkable water was scarce. Hotel Karibe—once a lavish conference venue—was still undergoing reconstruction. A dozen unknown scholars, myself included, slept on mattresses that covered the floor of one of the hotel's conference rooms. White sheets hung from the ceiling in an attempt to create some privacy. The room was crowded, noisy, warm, and smelly—and yet, it was probably one of the best accommodations in town.

The disaster killed thousands of people (estimates vary between sixty thousand to two hundred and fifty thousand deaths). The presidential palace was leveled, and the government was almost paralyzed. Hundreds of aid agencies had arrived in Haiti over the past months. Politicians in Washington, New York, Montreal, Brussels, and Paris had promised millions of dollars for reconstruction. The challenges ahead were enormous. Dozens of people had come to Port-au-Prince that warm day in June to discuss reconstruction objectives and to identify ways to achieve them. Local leaders, scholars, and journalists all wanted answers: What to prioritize? Where to start? When and *how* would it be done?

Local dignitaries, scheduled to speak in the morning, arrived late. They presumably had more important things to do. Scholars who spoke in the afternoon appeared to show more enthusiasm. When it was my turn, I explained the challenges faced in previous cases of postdisaster reconstruction and disaster prevention. I then argued that the top priority should be to upgrade infrastructure in the multiple slums of Port-au-Prince. The *cités*, as they are known, would improve if infrastructure was built and the informal sector was involved in reconstruction efforts.

This idea did not attract the interest of decision makers or the media. Their excitement surrounded the more ambitious reconstruction plans, unveiled to great fanfare in the afternoon. By then, local and international journalists were toying with more radical approaches. They noted that the city was built on a fault line, exposing it to tsunamis, tropical storms, hurricanes, and sea-level rise. Some argued that even before the disaster, the city was a total mess filled with impossible traffic, crumbling infrastructure, and poorly built housing. The idea that Port-au-Prince was a realm of total anarchy was taken for granted by most international experts, journalists, and even Haitians themselves. So they began to pose an intriguing question: Why not relocate the city altogether?

I was initially baffled. The idea of relocating a capital city of more than three million people seemed absurd. I imagined the journalists were just trying to get the hopelessly idealistic options out of the way first, and that soon we would focus on real solutions to the problem. But instead, geographers and global warming experts started to seriously discuss the idea of the city's relocation. They explained in the newspapers and on TV that a sustainable, brand new city should be built in a safe area. Survivors should be moved to new ecofriendly, solar panel–powered houses. Within a moment of what seemed to be postdisaster hysteria, the debate was no longer about whether the idea was plausible. It focused instead on how far the new capital should be placed from the coast in order to mitigate the effects of climate change–induced sea level rise. Total relocation was presented as an ideal opportunity to implement sustainable development at a regional, and even national, scale.

I thought this was yet another desperate attempt to cope with the trauma caused by the disaster. I expected that the folly would gradually subside. But there I was, astonished to see more than a dozen plans for Port-au-Prince's transformation drawn up within months of the earthquake. They all showcased variations of the city relocation idea: this neighbourhood, located here, will be transferred there. That market will be eradicated and replaced with this new highway. This district should not be here, where the soil is unstable and the heat unbearable. It would fit better there, where the soil is fertile and the breeze is cool. As for the cités, most plans prescribed demolition and replacement with more dignified housing.

THE ECO-CITY THEY WANTED

A few months later, colleagues from my research team and I were invited to moderate a panel at the Eco-City World Summit in Montreal.[2] The panel included Jean-Yves Jason, the mayor of Port-au-Prince at the time, who was in town to present his view of the reconstruction process. His plan for the capital city had been prepared by the IBI Group, a Canadian urban design firm.[3] In it, a great portion of the central district would be demolished and replaced with modern, energy-efficient buildings. The port would be expanded and made available to large cruise ships, while green boulevards would connect the new buildings for ministries and other public institutions (see figure 0.1). When I met Mayor Jason at the summit, he was convinced that he would transform the capital city in a sustainable manner. "This is the moment to create a new city and a new future," he told us. It became clear to me that, although the mayor had the drawings, he had few ideas for realizing this vision. There was no process in place for dealing with finances or convincing his opposition to approve of the plan.

Yet Jason was not alone in the search for radical change in Haiti. Within months, hundreds of Haitians and sustainability experts—from governments, international agencies, and charities—had eagerly planned for major urban transformations. The Clinton Foundation launched a sustainable development plan called Building Back Better (which I will explore again in chapter 6). The purpose was to build houses in the outskirts of the capital, in an area called Zoranjé, which would become a new model neighborhood. The project included a prototype housing exhibition and the development of "an exemplar housing settlement."[4] Bill Clinton explained, "In my 35 years of traveling to Haiti, I truly believe this is the moment that holds the most promise for Haiti to overcome its past, once and for all."[5] Advisors of the Clinton Global Initiative promised green reconstruction, predicated on the argument that "sustainable reconstruction can expand opportunity by investing in local entrepreneurs and supply chains." More promises, laden with sustainability rhetoric, followed. The idea was to drive recovery and reconstruction with sustainable energy: "Greening the recovery will also position Haiti to thrive in the coming low-carbon economy." This green recovery, the advisors claimed, "will punish wasteful resource use and reward innovation and human ingenuity in using resources in ways that yield higher quality economic and social outcomes."[6]

The Prince Charles Foundation also stepped in. The foundation's mission is about "respecting the past, building the future," and "supporting people to make the most of their community." Much like the Clinton Foundation, this organization is committed to "championing a sustainable approach to how we live our lives and build our homes."[7] But in Haiti, the foundation proposed a plan to demolish several blocks of downtown Port-au-Prince. Here, the idea was to replace the existing central district and slums with grand new green construction, plazas, and parks. This would require bulldozing

FIGURE 0.1 (*Top*) Architectural rendering of the Port-au-Prince reconstruction project, commissioned by Mayor Jean-Yves Jason. Plans included green boulevards and comprehensive port upgrades designed to receive multistorey cruise ships. (*Bottom*) A downtown charcoal market in postearthquake Port-au-Prince.

Sources: (*Top*) Haitian Government. Image of public documents disseminated by Haiti Libre: https://www.haitilibre.com/en/news-3558-haiti-reconstruction-3-3-billion-for-a-new-port-au-prince.html. (*Bottom*) Gonzalo Lizarralde, 2015.

several buildings with historic value. There would not be so much respect for the past when it came to building the Port-au-Prince of the future, it seemed.

These charities were not alone in trying to champion sustainable initiatives. Hundreds of architects, urban planners, environmentalists, and engineers seemed convinced that Port-au-Prince was a malleable product, to be molded in the way that an artist shapes a sculpture. They all had ambitious, but untested, green ideas for transforming the area in the aftermath of the disaster. "How could they all be so wrong?" I thought.

As time passed, my confidence began to wane. Could so many experts actually be mistaken? What if I was the one missing the point? By focusing on existing informal settlements, perhaps I had settled for mediocre solutions. Maybe I was failing to see the real opportunity—to achieve radical, real sustainable development in Haiti.

Six months after the earthquake, the Haitian government reported to the United Nations that only 2 percent of the promised aid had been effectively delivered.[8] The Center for Economic and Policy Research estimates that the U.S. Agency for International Development (USAID) awarded a total of $2.3 billion to Haiti. But more than 55 percent of that aid was given to organizations or companies located in the United States, with only 2.3 percent going directly to Haitian companies or organizations.[8] Years later, none of the radical, sustainable plans had come to fruition.

The Building Back Better relocation program never got off the ground, and the central district redesign proposed by the Prince Charles Foundation never moved forward either. The slums were not replaced with the wide boulevards and palm trees that decorated the planners' artistic renderings. The compact districts with their open sewers did not transform into the green neighborhoods envisioned by charities and agencies. The dilapidated industrial buildings were not razed to make way for the urban parks dreamed up by experts—and inevitably, no fancy cruise ships docked in the port.

Bill Clinton's visits to Port-au-Prince became less and less frequent (he became, I presume, more interested in his wife's race for the White House). The representatives from the Clinton Global Initiative, along with most international charities, left Haiti altogether after three years. In February

2012, political adversaries expelled Jean-Yves Jason from office over a series of minor scandals. (Jason blamed political persecution for his ousting.[9])

Meanwhile, corruption in Haiti remained rampant. In 2018 and 2019, scandals about the use of aid funneled through PetroCaribe (a funding program resulting from a cooperation agreement with Venezuela) led to violent demonstrations and strikes. The strikes largely paralyzed the country and economy.[10] Many observers claimed that the political system in Haiti was so severely compromised that it could hardly be saved. They even argued that the state was "locked" (they used the Créole term *peyi lòk*) and inoperative.[11] A "re-founding" of the Haitian state was necessary to save the nation from total chaos.[12] Local narratives focused on "renewal rather than recovery"[13] and condemned metastasized corruption,[14] patriarchal structures,[15] and "anarchic" forms of urban development.[16] But when I spoke to Haitians, I often found that they were troubled by how to distribute responsibility for risk creation, and unsure of how much responsibility to attribute to local corruption, foreign interference, and the legacy of racism and colonialism.

On my subsequent trips to Port-au-Prince, I was deeply saddened by the realization that little had improved in the country. Residents kept telling me that they felt deceived by the politicians who promised funds and sustainable development plans. Not only did politicians and international consultants break their promises of a green, successful future, they did it in times of profound suffering. In this way, agencies and governments—both local and foreign—had added insult to injury.

I must confess, however, that I was relieved by the failure of this radical change. I hadn't missed the point, after all. Postdisaster Haiti was never meant to be treated as a country-sized sustainable development experiment. Disaster reconstruction should not be the time to implement untested green ideas. But just as my confidence recovered, I heard about Canaan.

CANAAN, THE NEW HAITIAN MODEL CITY

A few months after the earthquake, about sixty disaster-affected families occupied a golf course in one of Port-au-Prince's wealthiest neighbourhoods. The tents that dotted the golf course were disconcerting to those in power. Since the earthquake, the American actor Sean Penn had been busy providing emergency aid to temporary camps in Port-au-Prince—and legend has it that Penn himself had met with the families to suggest relocating them to a region in the northern mountains of Port-au-Prince called Corail. Whether Penn actually convinced the squatters to relocate is still a matter of debate, but the fact is that families moved and soon occupied temporary units in Corail.[17] Even though there were no jobs, roads, or public services in the region, charities quickly built more temporary shelters. Within a few weeks, Corail became a magnet for other residents. As the area began to grow with more shelters, it became known as Canaan, a name that conjures the biblical promised land.

A few months after the disaster, a single signature changed the course of Canaan forever. René Préval, then the president of Haiti, signed a bill marking the territory of Canaan for public use. When Préval signed the bill, thereby expropriating landowners, he did not appear to consider how this change would affect the population. But in Haiti, land ownership is a matter of life and death. It is largely controlled by politicians, the elite class, and land cartels. People are routinely murdered in land tenure disputes. With the stroke of a pen, Préval made a tacit (some say deliberate) invitation for Haitians to occupy this territory.

A hundred families quickly became a thousand—then tens of thousands, then a hundred thousand. Today, Canaan is home to more than three hundred thousand Haitians. It is a brand new, functional city—probably one of the largest created in a period of only seven years. Canaan was neither planned by international consultants and official authorities nor built by corporations or well-known construction companies. Instead, Canaan was designed and constructed by ordinary citizens, community leaders, and real-estate mafias.

Canaan lacks the green boulevards that postdisaster experts sought to build in the capital. There are no tourist-ready promenades like the ones foreign planners envisioned for the city's waterfront. The streets aren't even paved.

Canaan is not (yet) recognized as a municipality. But it is no slum either. It has plenty of functional parks, churches, schools, and wells. Most infrastructure and services have been built by citizens, community leaders, and local organizations. Not everybody in Canaan is poor either. The place has a mix of low-income residents, middle-class families, industrial workers, and professionals. It also has a beautiful view of the Caribbean and direct connections to two national roads.

Land mafias have largely controlled the subdivision of land and selling of plots. The business is profitable. In just a few years, land prices in some areas of Canaan have multiplied twenty times over. Most construction is still precarious, but there are also hundreds of finished two-storey homes and operational businesses. In 2018, a rise in crime and violence affected the perception of safety that had existed in the early years of Canaan, yet most residents and local leaders stay optimistic about the future. They do not want Canaan to become yet another Haitian slum, and they remain convinced that they can still create a model city. Most roads have names that reflect this enthusiasm: rue de la Découverte, rue la Grâce, rue Prospérité. Even in 2020, Canaan continues to grow at a fast rate.

Canaan is tangible proof that I was wrong. Metropolitan Port-au-Prince had a capacity to change that I had totally missed. But it did not change in the way that experts anticipated. Nobody predicted that Canaan would become one of the largest settlements in the country. So after the Haitian disaster, when politicians, charities, academics, and international agencies all called for change, change did occur—it just was not the type of transformation that people in New York, Washington, Montreal, and Paris preferred. The mighty Goudougoudou triggered a chain of events that neither Haitians nor international donors and consultants could foresee.

Postdisaster reconstruction in Port-au-Prince exemplifies a trend in the world of unnatural disasters: major destruction, or the risk of it, is seen as an ideal moment for radical change and a chance for those in power to redesign from scratch. This is what journalist and writer Naomi Klein called "the blank slate" in *The Shock Doctrine: The Rise of Disaster Capitalism*. Her book, a 2007 bestseller that linked disasters to the implementation of neoliberalism and savage capitalism, is now viewed as a classic in disaster studies. In this book, I argue that implementing savage capitalism and neoliberal agendas for the benefit of corporations and foreign organizations is not always so direct. When it comes to risks and disasters, political and economic elites have found other ways to sweeten the pill of neoliberal capitalism—by sugarcoating it in narratives of sustainability, resilience, citizen participation, and innovation.

• • •

To be sure, the Goudougoudou was not totally extraordinary. Over the past decades, Haiti has been affected by many other tragedies. Hurricanes Hanna and Mathew hit the country in 2008 and 2016, respectively. A cholera epidemic followed the 2010 earthquake, and a period of significant social unrest plagued with looting, economic paralysis, and violence took place in 2018 and 2019. The International Displacement Monitoring Centers reports that, in the first half of 2019 alone, as many as 2,100 Haitians were displaced due to violence and disasters.[18] In fact, it has become difficult to stay up to date with news of disasters and social unrest in Haiti—a difficulty that is increasingly prevalent in other countries as well.

If you think disasters are on the rise in your own country, you are not alone. Munich Re, a think tank that records catastrophic occurrences, estimates that there were four times as many disasters in 2015 as in 1970.[19] Weather-related destructive events have tripled in the past four decades. In 2017, there were six times more destructive events caused by water than in 1980,[20] and in 2019, a global report found that "the number of people living in internal displacement is now the highest it has ever been."[21] Conflicts, violence, and extreme weather events are responsible for most of this displacement.

The economic consequences of disasters are also on the rise. In 2017, costs from disasters worldwide were ten times higher than in 1987.[22] It is estimated that in the United States, the number of extreme weather events causing at least $1 billion in economic losses has quadrupled since the 1980s.[23] In the past ten years alone, the California Department of Forestry and Fire Protection has spent over $3.8 billion in response to fires in the state, more than what it spent in the previous thirty years combined.[24]

To be sure, some natural hazards, such as floods, droughts, lighting strikes, volcanoes, and storms, cause fewer deaths today than they used to. But other unfortunate outcomes still occur—including injuries, displacement of people, and destruction of property—which may become even more widespread as global warming makes certain natural hazards more frequent and intense. The World Health Organization, for instance, estimates that global warming will cause two hundred and fifty thousand additional deaths annually between 2013 and 2050 due to higher rates of malaria, malnutrition, diarrhea, and heart stress.[25] A 2019 United Nations report on disasters concludes that "the way in which such changes—including in the intensity and frequency of hazards—affect human activity is as yet difficult to foresee."[26] Pollution, deforestation, and overexploitation of water, soil, minerals, fossil fuels, and land are exacerbating disasters everywhere. People and the environment have never been at higher risk—and development is largely responsible.

UNSUSTAINABLE DEVELOPMENT

Alarmed by the prospect of destruction, we coined a term that offered hope: "sustainable development." In 1987, sustainable development was elevated to the status of international policy in a report by the Brundtland Commission (a body established by the United Nations) entitled *Our Common Future*.[27] The idea of green, eco-friendly development quickly captured people's collective imagination worldwide. Green buildings, neighborhoods, and cities promised a way to maintain our lifestyle that could be passed on to future generations guilt free.

However, as disasters became more frequent and costly, the idea of resilience (a term as abstract and malleable as "sustainability") rose to the forefront. Viewed as an extension of the sustainable development agenda, "resilience" became a popular catchword for plans to deal with ongoing and imminent destruction. While the green agenda emphasized new ways to reduce impacts *on* the environment, the principle of resilience focused on how communities could avoid disasters triggered *by* it. Resilience became part of the vocabulary of journalists, theorists, and technocrats. In policy, it became the ultimate goal. What could be better than maintaining the status quo, uninterrupted, regardless of the impending disaster?

Trendy buzzwords like "sustainability" and "resilience" laid the basis for a multimillion-dollar industry of books, consultants, courses, conferences, accreditations, and awards. But despite the popularity of these ideas, annual CO_2 emissions worldwide grew by 1.7 times between 1987 and 2015.[28] This was enough to increase global temperature by about 0.4°C,[29] putting us closer to climate-induced Armageddon worldwide. We also lost about 2.5 million square kilometers of Arctic sea ice, which raised the sea level by 80 mm, making disasters all too common in coastal areas.[30] Wild species' populations declined by 60 percent, and almost half of shallow-water corals were lost.[31] Changes in the ocean caused by global warming might have "alarming consequences for global oceanic oxygen reserves," which have been reduced "by 2 percent over a period of just 50 years."[32] We have dumped so much plastic in the ocean that many of us are now inadvertently eating it.[33]

Disasters of all kinds are prompting unprecedented displacement and massive migrations worldwide. Hurricanes Katrina, Sandy, Harvey, and Irma caused thousands of deaths and billions of dollars in property damages in the United States. Between 1999 and 2018, 148 tropical storms and hurricanes killed more than two thousand people in the U.S. mainland.[34] In 2017, almost three thousand people died in Puerto Rico after the passage of Hurricane Maria. The 2004 Indian Ocean earthquake and tsunami displaced thousands of people, and an earthquake in 2003 partially destroyed the historic city of Bam in Iran. Every year, heat waves, floods, and wildfires displace populations and kill millions of people worldwide.[35]

What happened? Why couldn't we contain the destruction?

My answer stands on the shoulders of Naomi Klein's project, as presented in her disaster trilogy *The Shock Doctrine*, *This Changes Everything*, and *On Fire*.[36] Klein's inquiry aims to reveal alarming connections between different sorts of disasters (in Puerto Rico, New Orleans, Sri Lanka, South Africa, and Iraq, among others) and the free-market ideology promoted by the American economists and politicians of the so-called Chicago school. She applies the term "shock therapy" to the strategy of using disturbances caused by disaster and fear to enforce neoliberal fundamentalism and unregulated corporate action.

Neoliberalism entails political and economic practices that defend "strong private property rights, free markets, and free trade," and which often manifest in "deregulation, privatization, and withdrawal of the state from many areas of social provision." According to David Harvey, a professor at the City University of New York and author of *A Brief History of Neoliberalism*, this doctrine entails much "creative destruction . . . capable of acting as a guide to all human action" and substituting "all previously held ethical beliefs."[37] The connections Klein makes between this form of "creative destruction" and the use of violence in times of disruption are crucial to our understanding of contemporary politics and economic systems. Her work has opened our eyes to a series of compelling relationships between economic systems and disaster violence. In this book, I argue that many cases of risk creation—including recent disasters in Haiti and Central America—can indeed be linked to neoliberal policy and unrestrained capitalism. But having established these links, my inquiry leads me to other cause-effect relationships that are not at the center of Klein's project. As a matter of fact, research points me to five intersecting themes that you will find in next chapters and the stories presented in them.

INTERSECTING THEMES

First, the reasons why social injustices are reproduced during disasters go far beyond simple neoliberal fundamentalism. Corruption, stupidity, elitism, naïvety, ignorance, inattention, racism, colonialism, personal vendettas, and a sense of superiority all feature prominently in risk creation and the disaster management industry. To be sure, the desire for political and economic power is often the overarching context in which these crimes and injustices still occur. But there is so much more to our incapacity to stop disasters—and our capacity to create new ones—than a few powerful men trying to control the global economy.

Besides, we will find that when neoliberalism meets risk, it does not act as a single doctrine. It takes different forms and leads to a myriad of responses that must be explored from different angles. In the cases I present here, we will see the influence of common neoliberal agendas: rapid liberalization of markets, coerced integration in international trade agreements, government decentralization,[38] systematic transfer of responsibilities to markets and individuals, irresponsible reduction of public institutions, heightened control by corporations, business-government alliances, and other forms of savage capitalism.[39] But we will also see that economic policy feeds not only on the power of international businesses, geopolitics, colonialism, and imperialism but also on local ills such as municipal and national corruption, elitism, tribalism, and even inexperience and authorities' naïve desire to simplify complex problems.

Second, I have found that the disaster reduction and response sector is full of employees and decision makers who have no sinister plans in mind. Most are trying to act in good faith. Many accept that the problems they face are beyond their capacities to solve. And some are trying earnestly to transform the overall system from within, one small action at a time. But institutions are complex systems that do not always operate under the ethical principles held by the people working within them. In other words, the machine sometimes moves in directions that even the majority, pedalling from within, cannot control.

In this book, I explore not only the sinister effects of those acting with malice but also the unintended consequences, hidden impacts, and long-term secondary effects caused by people with good intentions. In many cases, failure to respond to people's needs and expectations can be traced back to the complexity that cities and human systems have achieved. This does not mean, of course, that I exonerate decision makers for their mistakes—my intention is the opposite. But it does mean that in order to produce positive change we must address the complexity of contemporary human systems and the diversity of interests that are at play.

Third, we must broaden the scope of responsibility. I argue that we have collectively failed to see that most disasters are, one way or another, the result of uncontrolled urbanization, poverty, marginalization, colonialism, racism, and other social injustices. We have ignored the fact that their effects are typically exacerbated by violence, consumerism, inequality, and the pollution we create every day. But again, this does not mean we can exonerate those who control the strings of political and economic action. Nor does it mean diluting responsibility into meaninglessness under the premise that we are all equally responsible. Politicians and decision makers, paying lip service to sustainability and resilience, like to make us think we are all responsible for global ills. In a way, we are, and we should recognize that—but we should also refuse to be complicit in the manipulation game that such slogans facilitate.

I particularly target four ideas in this book: sustainability, resilience, community participation, and innovation. I will explore their roles in our failure to tackle risk and destruction. We will discover, for instance, that for some, sustainable development has never been about protecting nature but about ensuring continued access to the resources needed to maintain economic growth. I will conclude that sustainable development has become unsustainable.

Fourth, I dwell on the importance of explanations, descriptions, and meanings. I will argue that we have adopted the language of risk but have ignored the risk of adopting this language. In this way, we have become complicit in depoliticizing climate change action,[40] reconstruction, and disasters.[41] As consumers, for instance, we have not noticed that sustainability and resilience have become a way to keep up business as usual while making a small effort to protect people and natural resources. In the meantime, sustainable development has become a profitable source of consulting revenue. Sustainability and resilience have been hijacked by economic and political elites to advance savage capitalism and shield themselves from criticism. They have become mirages—pop-culture products to manage individual guilt while providing a false sense of collective agreement.

Finally, I will contend that we have severed our understanding of how natural hazards affect people from our understanding of how humans affect the environment. In both scholarship and policy, human disasters have been disconnected from pollution and the destruction of ecosystems. Adopting an anthropocentric approach, we have understood disasters as major events that affect people, often forgetting that people have become the disaster that destroys nature. We have failed to see that unprecedented transformation of landscapes and the loss of biodiversity are also unnatural disasters in their own right—even when they do not directly (or immediately) affect people's lives.

These are the central arguments of this book.

The causes and consequences of these problems are interwoven. Much like the inhabitants of Borges's Babylon, we are gambling with our future—we are playing an infinite game of chance. When it comes to prevention, we disagree on what sacrifices should be made to protect nature. Most of us are still unwilling to give up economic growth, traveling, and consumption in the name of environmental protection. We disagree on how (or even whether) authorities should protect citizens. Risk reduction plans are frequently met with opposition. Tensions mount between environmentalists and governments, and political deadlock is increasingly common, stalling the most crucial reforms.

When disaster strikes, most responses are weak. Temporary solutions often become permanent, trapping beneficiaries in substandard living conditions. Interventions are fragmented, leading to both redundancies and omissions in the aid provided. Solutions rarely meet people's needs or expectations. Vulnerabilities are often replicated.

This book attempts to avoid oversimplifications, instead exploring contemporary debates about the relationships among cities, natural hazards, and savage capitalism. It draws a portrait of society based on the losses we have already witnessed and the ones we might endure in the near future.

Throughout, I present postearthquake Haiti and other cases in Canada, the UK, Italy, the United States, Vietnam, Colombia, South Africa, Honduras, El Salvador, Chile, and Cuba as examples of the numerous injustices that exist in a world of unnatural disasters, and I also reveal the efforts made around the world to halt climate change, prevent disasters, and rebuild after them. You will find arguments from Bill Clinton, Silvio Berlusconi, Narendra Modi, Enrique Peñalosa, and other politicians leading efforts to avoid disasters or respond to them, in addition to the hundreds of people I have interviewed or worked with over the past two decades. You will hear the voices of disaster-affected people resisting rapid transformation and unrestrained capitalism in Chile, Colombia, and India, among other places. You will encounter virtuous environmentalists and activists engaged in reducing carbon footprints and emissions in the UK and Canada. You will meet Central Americans escaping disasters at home and finding new risks abroad. Finally, you will meet ordinary citizens (mostly women) producing small changes in their own communities in Latin America and the Caribbean.[42] To be sure, most men in power are ignoring their actions—but today, the unpretentious activities of these local leaders constitute our best hope.

These are the stories and the science of unnatural disasters.

CONNECTING THE DOTS

This book is largely about making connections among seemingly unrelated events. The connections are unpacked in nine chapters.

The first chapter deals with the causes of disasters. Destruction happens for a reason but not the one we tend to think of. In chapter 1, I argue that the so-called refugee crises in rich countries cannot be fully explained without a thorough understanding of the hazards and vulnerabilities in poor nations. I explore the interweaving events and decisions that lead to displacement and devastation in Central America, and I show how disasters have become the causes and consequences of some of the most pressing challenges societies face today. This case helps me introduce academic debates about the real causes of disasters, showing that scholars still disagree on the extent to which disasters are caused by natural or human factors.

Chapter 2 is about change—the need for and consequences of it—following disasters. Here, I explain how disasters and the prospect of destruction are often seen as ideal moments for modernization and radical transformation. I also explain how environmental protection, risk reduction, and reconstruction are always political. As such, they produce winners and losers. In this chapter, I introduce the debate around the benefits and drawbacks of aid, showing how most analysts resort to abstract generalizations and assumptions about the role of charities and humanitarian action. I explore the types of change that people in power often want and conclude that in the face of risk, authorities, international consultants, and foreign charities are often disconnected from people's expectations and urban realities.

Chapter 3 focuses on the concept of sustainability. Most citizens and professionals have gotten behind the idea of sustainability in efforts to do what they believe to be the right thing. But even though most people recognize the importance of protecting nature, we do not necessarily agree on how to achieve it or what price to pay for it. In this chapter, I contend that sustainability is a useful tool to mask those differences and make us believe that we are all in the same boat. Unsurprisingly, sustainability has become a popular cloak for greenwashing and other facile marketing strategies. Here, I explain how difficult it truly is to protect ecosystems and reduce pollution—even for the most committed environmentalists. If we are serious about protecting nature, we need to make significant sacrifices and stop pretending that our symbolic gestures are enough. We need to do more—and better.

In chapter 4, I deal with the advantages and disadvantages of the resilience framework. I explore how households and businesses are increasingly expected to *be*, or to *become*, resilient in a world afflicted by unnatural disasters. The principle of resilience was originally adopted to show how individuals and groups could avoid disasters and reduce risk. Even today, there are numerous well-intentioned professionals and decision makers promoting resilience in attempts to reduce risk. But politicians, charities, corporations, and consulting agencies worldwide have also found a way to capitalize on the idea of resilience to advance more sinister agendas. I illustrate how authorities and urban consultants worldwide have hijacked the resilience narrative and how the concept has become a tool to legitimize government disengagement with the most vulnerable, to maintain the status quo, and to carry out projects that predominantly serve the interests of economic and political elites.

Chapter 5 is a (sure to be controversial) discussion of people's participation in disaster risk prevention and reconstruction. Today, millions of citizens spend time in participatory meetings and consultations because—as almost every politician, NGO officer, consultant, and design professional will tell you—the key to success in disaster response is community participation. Participation is also one of the most commonly discussed topics in the fight against climate change, but is it actually the key to success? In this chapter, I assert that it is not. Participatory activities have, for the most part, become an extension of sustainable development rhetoric, and no matter how heartwarming they appear, they rarely meet the needs and expectations of ordinary citizens and local leaders.

Chapter 6 also raises a controversial issue—the role of innovation in the face of risk. Here, we meet both the winners and losers of radical innovation in times of disaster. Massive destruction is often followed by a set of enthusiastic engineers, architects, designers, and part-time innovators proposing their latest "inventions." These include shelters made of containers that can be "deployed to disaster-affected areas in just a few days," floating houses (and even cities) that can literally navigate the threats caused by climate change, inflatable structures that can "simply" be airlifted to remote areas by helicopter, and other idealistic contraptions. Here I argue that most of these "solutions" fail to solve the real problems faced by disaster victims; many of them create new problems. But, of course, high-tech dreams do not just erupt after disasters. There are scores of scientists and ordinary people hoping that technological discoveries will help us halt climate change and protect nature. Critics often respond that it is precisely this technology that is moving us closer to planetary Armageddon. Meanwhile, defenders of high-tech discoveries celebrate—and their critics deplore—that technological innovation is the easiest way to maintain economic growth.

Is there an alternative to the sustainable development narrative to counter disasters and the risks that they pose?

Having explained the narratives of sustainability, resilience, participation, and innovation in chapters 3 to 6, I propose in chapter 7 an exploration of alternative narratives and explanations of risk, disasters, climate change, and development. By focusing on the type of leadership that is needed to face global warming and disasters today, I challenge common premises. For instance, I demonstrate how it is often better to reinforce existing settlements and slums than to replace them with new housing—and additionally, I explain why disaster victims often prefer to receive cash instead of houses. Here I show examples of approaches that have refused to embrace the narratives of sustainability, resilience, participation, and innovation, and have instead assumed an ethical approach to risk.

In the final chapter, we come to terms with humility. You will encounter community leaders on their respective missions to halt deforestation, reinforce fragile housing, and develop income-generating activities in their own communities. These cases serve as a background to explore the values required to redress social injustices and save a planet at risk.

I conclude by focusing on ideas targeted to practitioners, academics, and citizens in general. I will stress that as citizens, we must learn to distinguish superficial solutions from real ones. We must also recognize that many green endorsements, building certifications, and fair-trade labels have become sophisticated marketing schemes rather than effective solutions to reduce pollution. As scholars, we need alternative narratives of risk, disasters, global warming, and development. We need to pay more attention to voices from the margins and amplify them to reconstruct new (more appropriate) explanations. New approaches must be based on principles of social and environmental justice, and they must recognize the role of people's emotions, knowledge, meanings, and values. Finally, as practitioners, we must recognize the importance of bottom-up decision making in the face of both risk and destruction. I conclude that protecting the environment and ensuring the safety of human beings are no longer two different things.

Throughout eight chapters, I develop a narrative arc in which I invite readers to transcend their search for the "ultimate" green, resilient, innovative, or participatory solution. I implore them to realize the enormous power that we hold in determining our way of life in a post–sustainable development era. I contend that the quality of our society is measured not by our capacity to adopt the narratives of sustainability, resilience, participation, and innovation but by the way that we treat our most vulnerable members. This demands an unequivocal engagement with—and empathy toward—marginalized and neglected individuals and social groups. Contrary to what most international consultants and academics lead us to believe, I suggest that there is no magic solution or step-by-step guide to overcoming these challenges. In a world of highly interconnected problems and complex systems, there is no easy way out.

Good intentions are not enough to rescue people, animals, trees, and oceans. The solutions we need today are expensive and difficult. In a post-sustainability age, we will all have to make real sacrifices; not the symbolic ones we are accustomed to. These sacrifices will be felt immediately at the individual level, whereas the rewards will be reaped collectively and only in the long run. It won't be easy . . . but we have little choice.

1 | Causes

"Disasters Happen for a Reason"

A Babylonian is not highly speculative. He reveres the judgement of fate, he hands his life over to them, he places his hopes, his panic terror in them, but it never occurs to him to investigate their labyrinthian laws nor the gyratory spheres which disclose them.

—Jorge Luis Borges, "The Lottery in Babylon"

RODRIGO CHINCHILLA AND THE RECONSTRUCTION THAT NEVER HAPPENED

In fall 2018, a caravan of people from Honduras, El Salvador, Nicaragua, and Guatemala walked through Mexico on its way to the United States. For years, Central Americans in similar caravans have sought asylum in North America, but this was the largest reported to date.[1] Some journalists counted as many as seven thousand migrants.[2]

The walkers endured countless difficulties. They had few belongings and little food. They walked in harsh weather and crossed rivers, city centers, and highways by foot. They faced violence, hostility, tear gas, and military barricades in Guatemala and Mexico. And when the group temporarily settled in parks and public spaces in the towns they passed through, they were met with resentment and anger.

That's when President Trump jumped in to make a political case against the walkers (and more generally, against what he calls "shithole countries"). He had previously called Latin immigrants in America "opportunists" and "rapists,"[3] before making the absurd claim that terrorists were hiding in the caravan. At that point, he sent the army to prevent the group from crossing the border.

As the migrants walked through Mexico, American troops waited for them in Texas.[4] A series of tragedies related to migration surrounded the event. Migrant children were detained in the United States and separated from their parents.[5] Two of them died while detained. Then, the U.S. government began its longest shutdown in history over a dispute between the White House and Congress regarding this so-called border crisis.

Images of migrants carrying their children on their shoulders shocked viewers worldwide. Many in the caravan were young children, as well as single parents who had been victims of violence or had lost their jobs.

Commentators tried to make sense of the tragedy. They wondered how difficult the conditions in El Salvador, Honduras, Nicaragua, and Guatemala must be for parents to put their children through this torment. Most analysts pointed to the political instability, poverty, and drug-related violence that plagues Central America.[6] However, few linked this tragedy to previous disasters.

The Central American isthmus is one of the most disaster-prone areas in the world.[7] The region sits on fault lines that cause frequent seismic activity, triggering earthquakes, tsunamis, mudslides, and landslides. It is also a densely populated area studded with active volcanoes—three of which erupted between 1999 and 2000 alone.[8] Its eastern coast lies in the path of tropical storms and hurricanes. Its western one is affected by meteorological events like El Niño and La Niña, which often intensify seasonal weather patterns. The region thus oscillates between harsh dry periods and torrential rains; human-induced climate change, pollution, and deforestation only exacerbate these effects.[9] Along with Haiti, Central American countries are at the top of the Climate Change Vulnerability, Exposure, and Sensitivity indexes produced by the Development Bank of Latin America.[10]

In Central America, politicians and geopolitics linked to American influence during the Cold War have created conditions for a chain of unnatural disasters. This brings us to a discussion of Rodrigo Chinchilla.

When I met Rodrigo in El Salvador's capital of San Salvador back in 2003, he had just returned from Canada after earning a master's degree in urban planning. Two years before that, El Salvador was hit by two severe earthquakes within just two months. About 1,259 people died and more than 150,000 homes were left in ruins. The disasters also destroyed 23 hospitals and 121 healthcare centers, while damaging 1,566 schools (roughly 40 percent of hospital capacity and 30 percent of the schools in the country). The total economic loss was estimated at $1.2 billion.[11]

A few months after the earthquakes, San Salvador introduced a program to build six thousand houses and several infrastructure projects.[12] I knew Rodrigo was in charge of the city's planning department, and I felt that if anyone was capable of transforming the city, it would be him. So I set up an appointment with Rodrigo and scheduled an entire day to meet with his staff and learn about their reconstruction projects.

Then reality hit: my visit took less than two hours. Rodrigo wasn't leading the planning department—he *was* the department. As the only municipal employee dealing with planning and development, there was nothing to see in his office besides a desk, his computer, a plotting machine, and a few maps.

Even more alarming was that Rodrigo had no reconstruction projects to talk about. He was a hardworking, responsible city officer, but a few years after the disaster, no permanent houses (and almost no new infrastructure) had been built.

The national government was still trying to construct temporary housing, but resources were scarce. Disaster victims were given pieces of wood and sheets of corrugated metal to build their own temporary shacks in a village called Tonacatepeque, a one-hour drive from San Salvador. But without infrastructure or public services, Tonacatepeque had become one of the most dangerous areas in the country. When I asked Rodrigo whether we could visit the new settlement, he shook his head, "Not even the police go there these days." So I asked him about other reconstruction initiatives. "The story I can tell you," he said, "is the story of the reconstruction that never happened."

What happened—or rather, why had *nothing* happened in San Salvador after the 2001 double disaster? Like many stories of human suffering, the answers to these questions start with violence.

The list of disasters in El Salvador, and in Central America more generally, runs parallel to a history of civil wars fuelled by corruption and Cold War politics. In the 1980s, a violent civil war erupted between the leftist guerrilla Frente Foribundo Marti para la Liberación Nacional, or FMLN, and the Salvadoran government. The war left about seventy-five thousand dead (a staggering loss in a small country of 6.5 million), with countless people missing.[13] At one point, the FMLN gained enough public support and military power to challenge the stability of the state. For more than five decades, the United States had worked fervently to prevent socialist movements from spreading in Latin America and to support politicians willing to accept economic conditions imposed by Washington. American politicians were not about to make an exception for El Salvador, so they dispensed economic and military support to the Salvadoran right-wing government (aid estimated at about $4 billion).[14] After twelve years of bloody combat, both the government and FMLN's forces were weak and worn out. In 1992, they signed a peace agreement.[15]

But American dollars, military support, and international loans came with political pressure to "modernize" Salvadoran public institutions.[16] The package of reforms that was prescribed to El Salvador—and other countries in the region—was a mix of Thatcherism and Reaganomics. The neoliberal cocktail aimed at significantly reducing the size of the state, increasing government efficiency, and decentralizing decision-making power.[17] This entailed transferring the responsibilities that the government had previously assumed to the (presumably more efficient) private sector and markets.[18]

At the time, a common neoliberal agenda was to eliminate ministries and agencies that had traditionally dealt with low-cost housing, infrastructure, planning, and construction.[19] The quest for "efficiency"—prescribed by countries in the Global North—included forcing governments in Latin America to privatize state companies (such as public utilities), reduce their public spending, and encourage private businesses to deliver services and infrastructure.[20] In exchange for aid, governments in El Salvador and other countries were encouraged (or forced) to stop providing direct public services, housing, and urban infrastructure.[21]

In many cases, several responsibilities were transferred to municipalities that often lacked the administrative, economic, and legal resources to deal with overwhelming deficits in the best of times.[22] Needless to say, they were largely impotent in the face of the exceptional needs that came with earthquakes, floods, and hurricanes. In 2001, for instance, a relatively small local charity called Fundasal built more houses in El Salvador than the national government and the city of San Salvador put together.[23]

Rodrigo was lonely in his small office. But he was by no means the only municipal officer in an understaffed and ill-funded department. A few weeks after meeting Rodrigo in 2003, I travelled to a small, yet historic, city in Honduras called Choluteca. The city was partially destroyed by Hurricane Mitch in 1998, and there I found even fewer resources devoted to housing, infrastructure, and services. The municipality was unable to cope with regular or disaster-related projects.[24]

In the past five decades, El Salvador, Honduras, and other countries in Central America experienced significant demographic growth and rapid rural-urban migration, which only enlarged their deficits in services, housing, and infrastructure. Slums proliferated.[25] Scores of poor and rural migrants settled among disaster-prone riverbanks, urban fringes, and hillsides. Without the skills required for jobs in the city, many rural migrants found themselves unemployed, while subsidies, welfare, and social programs shrank or disappeared. Corruption and neglect took a toll on the most vulnerable.

The region became a middle passage for drugs produced in South America and consumed in the United States. Without services, schools, or opportunities, many young men joined drug cartels, guerrillas, gangs, and other criminal organizations. Those who had escaped rural violence by moving to cities like San Salvador, San Pedro Sula, Tegucigalpa, and Choluteca fell prey to new forms of crime and aggression. Gangs and cartels metastasized in urban centers and slums.[26] Within just a few years, cities in El Salvador, Honduras, and Guatemala had some of the highest murder rates per capita in the world.[27]

Political instability and corruption certainly played a role as well. U.S.-supported military coups in the region were followed by state-led violence and oppression. Notorious dictator Efrain Rios, who became president of Guatemala in 1982 after a military coup, was later convicted of genocide by a Guatemalan court.[28] The civil war in Guatemala, from 1960 to 1996, left roughly 200,000 deaths and an unknown number of people missing.[29] In 2009, President Zelaya was ousted by a military coup in Honduras.[30] In 2011, Daniel Ortega violated the Nicaraguan constitution to begin his third presidential term.[31] Even during periods of relative calm, tensions between right- and left-wing groups routinely led to political deadlock. The left-wing administration that hired Rodrigo, for instance, was systematically boycotted and oppressed by the right-wing government that attempted to build Tonacatepeque.[32]

Not surprisingly, the natural hazards in the region triggered major disasters. Between 1970 and 2000, there was an average of thirty-two disasters per year in Latin America and the Caribbean, leading to 226,000 fatalities.[33] Between 1980 and 2017, about eighteen thousand people died in disasters in four countries alone: Nicaragua, Honduras, Guatemala, and El Salvador. During that time, Honduras had about twenty-five times more fatalities per capita (caused by catastrophic events) than countries like Cuba, Ireland, or Canada, and five times more than the United States or the UK.

Disasters in Central America not only resulted in thousands of deaths but also significant economic losses.[34] It is estimated that Hurricane Mitch alone caused direct and indirect damages of $6 billion. It also caused a 4 percent rise in poverty in Honduras (another disaster in 1993 had caused its GDP to contract by 3.4 percent). In Nicaragua, Hurricane Joan (1988) caused damages of more than $1 billion, and the droughts that followed in 1994 further reduced the country's GDP by almost 9 percent. Hurricane Mitch, meanwhile, caused Honduras's GDP to fall by 32 percent in 1998.[35] In 2000, economists from the Inter-American Bank concluded that these and other major disasters in Central America had led to substantial decreases in investment, exports, private savings, and debt payments, as well as increases in the deficit for balancing payments and government debts.[36]

Can we disentangle the displacement of Central Americans—and the humanitarian crisis that followed—from the disasters we usually call "natural"? And can we distinguish these disasters from other (perhaps less natural) disruptions, such as poverty, corruption, environmental damage, and violence?

I think we cannot—and what's more, we *should* not. The caravan migrants who met the U.S. soldiers at the Mexican border were indeed escaping violence, poverty, and corruption, as analysts aptly pointed out. But they were also escaping numerous disasters triggered by a combination of natural events and political decisions. Explaining the reasons for this outstanding displacement without referring to the *un*natural disasters that beset Central America portrays an incomplete picture. Examining those disasters instead provides us with a valuable lesson for the future: if risks are not reduced in the region, we can only expect more caravans to arrive at the U.S. border.

RIPPLE EFFECTS

The subtitle of this book references disasters and climate change. So you might be wondering: Is this book about the type of disasters we commonly call "natural"—or is it about the effects of human action on the environment, such as climate change and pollution?

We tend to believe that "natural" disasters, violence, and climate change are three distinct sets of problems. Victims of "natural" disasters probably make you feel compassionate; you might think of ways to better protect people from volcano eruptions, seismic activity, floods, or storms. Many people also feel empathy when they recognize the hardships lived by refugees and displaced populations, but they commonly associate this suffering with poverty, crime, wars, genocides, and other forms of violence and social injustices. Finally, as we continue to receive updates about the ever-worrying rising temperature of our planet, we tend to think about harmful gasses and polluting activities. This prompts a different kind of feeling. We might be compelled to drive and consume less, plant more trees, and optimize machines, among other actions.

Decision makers and scholars typically deal with these problems in different ways. Organizations, for instance, fulfill very narrow missions; they rebuild after disasters, protect children, or save endangered species. Decision makers often have specific agendas in mind. Ministries and public departments, for example, rarely deal with problems as diverse as economic growth and disaster risk prevention. Finally, scholars are inclined to pursue focused areas of expertise, such as risk assessment, pollution, and disaster risk reduction. Yet, as the caravan case shows, these are no longer three sets of distinct challenges. Instead, the causes and consequences of current disasters are increasingly interconnected.

WHY WAS THE WOMAN KILLED BY THE TSUNAMI?

Sometimes the relationships between different damaging forces are easy to spot. Global Witness, an environmental charity, estimates that in 2017 alone, more than 197 activists were assassinated worldwide for opposing initiatives that threatened land, wildlife, and ecosystems.[37] In 2018, at least twenty-four environmental militants were killed in Colombia.[38] Put simply, people are killed *because* they seek to protect the environment.

Meanwhile, in 2019, heavy rain brought by Storm Norma led to flooding in more than 360 settlements and camps in Lebanon, affecting 11,300 Syrian refugees. More than six hundred Syrian refugees in the Lebanese region of Bekaa had to be relocated because of these floods.[39] Here, victims of violence become *the same* victims of a natural hazard.

In both cases above, the connections between tragedies are relatively straightforward. But in many other circumstances, these links are difficult to establish. Revealing them requires thorough consideration over longer periods of time, as they go beyond obvious cause-effect relationships.

It is easy to understand that fault lines under the ocean cause seismic activity, which displaces large volumes of water and caused major waves, which, in turn, cause floods and affect people in coastal areas. But it is more difficult to decode the underlying factors that lead to, say, a woman being drowned by a tsunami in a certain city.

Why does she move to that location? Why doesn't she receive warning? Why isn't she evacuated? What leads her—and not her brother—to assume the responsibility of taking care of her elders? Why is her house so structurally weak? Why is she working from home that day? Why doesn't she receive immediate medical attention?

Only in answering these and other similar questions can we truly understand the causes of unnatural disasters.[40] Yet, as in any other forensic job, the more interrelated the events are, the more difficult it is to understand why they happen.

AN UNLIKELY PARADISE

Imagine a society where all residents have universal healthcare and free, high-quality education. People live in small- and medium-sized cities. There are a limited number of cars on the streets; most people walk or cycle, while others (literally) use horsepower. Violence is rare.

In this world, there are no billboards marketing soft drinks, clothes, or any other product of consumption. Radios and TVs do not broadcast advertisements. Instead, they air educational programs about history, literature, sports, and music. Bars and cafés feature live music as well. People dance in the streets and attend free concerts, while children have free access to after-school activities. Baseball games are played on every corner.

Perhaps most notably, carbon emissions are extremely low. People consume only the essentials, and almost nobody wastes electricity. Other greenhouse emissions also remain low, as there is no overexploitation of cattle—or anything else. People rarely eat red meat. Machines and appliances are recycled and repaired. Low-carbon construction methods are the norm.

This is neither a thought experiment nor a green community in Scandinavia. This is Cuba. And yet, it is not paradise. Millions are desperate to leave due to widespread poverty, which is directly connected to politics—both local and global. In an effort to put pressure on the Cuban government that has ruled the island since Fidel Castro's revolution, American leaders imposed a commercial and political embargo on Cuba almost sixty years ago. Living on a small island isolated from international commerce, Cubans produce very few things. Diapers, clothes, computers, internet, sunscreen, appliances, construction materials, and electric tools are expensive and in short supply. The average income of Cubans is dramatically low. GDP per capita is US$6,445, or about one-tenth of the United States.[41]

Unrestrained consumption in Europe and North America—and now some emerging economies such as China—has become one of the most significant causes of environmental degradation and climate change. The remarkably frugal Cubans produce (in absolute terms and per capita) very little of the carbon and methane emissions that are contributing to these effects. Yet, ironically, they are on the frontline of climate change impacts. In fact, the island lies directly in the path of dozens of hurricanes and tropical storms that sweep the Caribbean each year.

Up until a few years ago, these tropical storms and hurricanes caused very few casualties in Cuba. These events routinely kill thousands in Haiti, the Dominican Republic, and Puerto Rico, among other islands, and result in millions of dollars in damages in the southern United States. The reason for the relatively low impact on human lives in Cuba is that, for the past few decades, the government has implemented dozens of initiatives to keep citizens safe.[42] In Cuba, a country with a very high level of education, climate change policy is driven by science and taken very seriously.[43] Risk prevention is taught in schools and universities; there are strong institutions devoted to risk management. The health system is strong and available to everyone. Information, risk communication, and early alert systems are accurate. Evacuations are timely and efficient.

All of these measures are implemented with the same type of authority that has kept a single-party political system in place for decades. But it is fair to say that for all its lack of financial resources, Cuba has strong social and institutional capital that makes it less susceptible to suffering from disasters. The problem is that climate change and variability—caused mainly by other countries' activities—are increasing these risks in Cuba. And with growing risks come difficult decisions.

THE FISHERMAN'S DILEMMA

Imagine you are the mayor of a small coastal city in Cuba. You have dealt with many threats in your life—from an American invasion to energy shortages—but now natural hazards are becoming more frequent and violent. The dry season has become drier, leading to wildfires and lost crops, and the rainy season is increasingly wet, which accelerates erosion and floods. The ocean has risen to cover land that was dry only a few years ago.

Among your responsibilities as mayor, you must guarantee the security and efficiency of the public facilities and infrastructure that provide services to Santa Maria, a small coastal fishing community of around six hundred people.[44] Climate change experts anticipate that half the village will be underwater within fifty years. If sea level continues to rise at current rates, Santa Maria will be totally submerged one hundred years from now (see in figure 1.1 images of coastal villages in the Santa Maria region).

You have two choices:

1. Let people stay in Santa Maria and try to mitigate the risks.
2. Relocate the residents to La Placita, a small village located a few kilometers inland with public facilities and infrastructure already in place.

Which one would you choose? Which one *will* you choose?

This scenario is not hypothetical; it is a reality faced by the decision makers I met in Cuba in 2017. If you choose the first option, consider the risks ahead. The cost of disaster prevention has already diverted funds from other pressing needs, such as education, security, affordable housing, and healthcare. In 2018 alone, the Cuban government had to increase its spending on housing and support for disaster victims by 50 percent, as a response to Hurricanes Irma in 2017, Matthew in 2016, and Sandy in 2012.[45] Caribbean nations are not alone in this respect; experts anticipate that the amount of resources needed for temporary relocations, periodic reconstruction, seasonal infrastructure retrofitting, and other accommodations will continue to increase in both developing and developed countries.[46]

If you choose the second option, you may have more in common with Cuban authorities than you probably thought. Today, national policy in Cuba prevents the construction of new housing in disaster-prone areas.[47] Families whose housing units are destroyed by hurricanes or tropical storms must relocate to safe zones. Reconstructing infrastructure in disaster-prone areas is also banned.

FIGURE 1.1 Coastal villages and fishing communities in Cuba are increasingly threatened by the effects of climate change.

Source: Gonzalo Lizarralde, 2016.

But your town is hardly the only one to adopt these radical measures. In fact, there are about 210 human settlements in Cuba located less than 1 kilometer from the water, including the major cities of Havana, Santiago de Cuba, and Cienfuegos. It is estimated that some form of relocation will be needed in almost all of these settlements in the next twenty years due to the effects of climate change.

In several towns, relocations have already begun: families are given apartment units in four- to five-story high buildings, often built with prefabricated construction systems. These new developments are located a few kilometers from the original locations (see figure 1.2). The apartment units have water and electricity and are relatively close to schools and infrastructure. Despite these benefits, though, many residents of Santa Maria are refusing to leave.

FIGURE 1.2 Members of a fishing community were relocated to this residential development in Cuba.

Source: Gonzalo Lizarralde, 2016.

AN OCEAN A FEW INCHES HIGHER

Manuel Cruz has seen much in his life. As a teenager, he experienced the last days of the U.S.-backed regime under Fulgencio Batista, followed by the tumultuous years of the revolution that brought Fidel Castro to power. He was there for the tense days when war with the United States seemed imminent; the years of relative prosperity, backed by Soviet aid; and the lean times when that nation collapsed. He knew the soothing relief of Venezuelan oil, and the return to desperation when that country also broke down. Then there was the promise of integration, peace, and tolerance upon the arrival of the first black American president, and the dissipation of that promise, accompanied by the sting of insults, when his successor took office. Mr. Cruz has seen the ocean come and go. He has seen nature's rage and beauty. I should have suspected that an ocean a few inches higher would not frighten him, and yet, in an interview in Cuba in 2018, I found myself asking him why he refused to be relocated to La Placita.

"Why should I go?" Mr. Cruz replied, without interrupting his chores. He was carrying three fish he had just caught and was preparing to clean his fishing equipment.

"Your town will be underwater in fifty years," I replied. "Besides, you could get a safe apartment in the new town."

"I am eighty years old," he interrupted. "Do you think that I am worried about the water? I will tell you something: the ocean has never harmed me. People have."

Mr. Cruz is part of a group of villagers who refuse to be relocated. For most residents of Santa Maria, their livelihoods come from the ocean. They fish, trade sponges and crab, take tourists out to admire the sea, and cook and sell delicious seafood dishes. Their fathers were fishermen and so were their grandfathers. The sea is not only a source of food and income: it is an essential part of their existence. Living in an apartment block far from *their* sea—even if it has running water and electricity—is, for them, simply ludicrous.

In other coastal villages some families have already accepted the apartment units, but the flats remain empty for a good part of the year. Many resist permanently abandoning their shacks and makeshift houses close to the sea. Of those who have relocated, many become depressed and require some form of psychological assistance. In Santa Maria and other coastal villages in Cuba, residents are more frightened of the apartment blueprints and relocation decrees from La Havana than the rising ocean.

The fisherman's dilemma exemplifies why people continue—or have few alternatives but—to live in dangerous areas. It also points to the difficulty of making decisions in the face of risk.

To be sure, we are all confronted by threats: our house being flooded, a fire at our workplace, or dying in a car accident. We constantly adopt different attitudes and responses to these threats—from buying insurance to saving money. We exude different "behaviors" to threats and catastrophic events, and we base our reactions on what people tell us about these harmful events.[48] Key factors include how the risk is presented to us, how much we trust the people charged with protecting us, and how likely we think it is that the risk will materialize.[49] We also assess potential losses, our familiarity with the event, and how much we can do to mitigate the risk—and at what price.[50] In some cases, we may even embrace and look for risk, notably if we see opportunities that we can exploit.[51] Building and running a guesthouse in Santa Maria is a risky business, for instance, but the profits lead dozens to assume the risk.[52]

While assessing risk, our minds play tricks on us. Sometimes we believe that negative events are more likely to happen to others than to ourselves.[53] Additionally, we tend to overestimate the risk that recent harmful events will happen again, and we underestimate the risk of events that haven't happened in a long time.[54] These cognitive tricks affect our decisions in the current, built environment.[55] Perceptions of danger also vary significantly among people. Authorities, for example, might see earthquakes and climate change–induced hazards as the primary risks to be tackled. Residents of informal settlements and historically-marginalized communities have a different perception of risk. They are often more concerned about daily struggles, such as unemployment, crime, food insecurity, and lack of water and infrastructure, rather than a possible disaster.[56] Cities and human settlements are shaped according to these individual and collective perceptions of danger and opportunities.

Citizens living in disaster prone-areas—in California and Athens, as well as Cuba and Colombia—know that disaster will strike but underestimate the likelihood that it will happen to *them*.[57]

The Cuban case helps us understand why low-income and marginalized people are exposed to hazards, but it fails to explain why both unprivileged and privileged people suffer from them. As we shall see, this is a controversial issue in itself.

"IN THE FACE OF NATURE'S RAGE, HUMANS ARE NOT ALL EQUAL"

"A divine chastisement for the great sin we have committed, and are still committing, against those whom we describe as untouchables."

These are the terms that Mahatma Gandhi presumably ascribed to the earthquake that occurred in Bihar, India, in 1934.[58] This scientifically unsound statement captures the essence of how people have explained disasters for centuries. Even today, many people still believe that earthquakes, fires, floods, tsunamis, and other tragic events are the results of supernatural punishments for human sins.[59] In 2014, Susanne Atanus, an American politician, claimed that tornadoes in the United States were a punishment from God after legislators' recognition of gay rights in her state. (Ms. Atanus won the GOP primary for the Ninth Congressional district of Illinois.[60]) Similar absurdities have been recently proclaimed in contexts as dissimilar as Brazil and China.[61]

Intellectuals have, of course, challenged the relationship between disasters and divine punishment for a long time. But in the 1980s, a group of scientists found that most scholars and technocrats had shifted the focus on natural events themselves, their characteristics, frequency, and intensity. The scientists also noted that many colleagues and politicians were more interested in the mechanics of planning and managerial activities to contain disasters, rather than the sociopolitical and economic conditions that lead to damage, disruption, and pain. In 1983, K. Hewitt, a professor of geography and environmental studies, deplored that "in the dominant view . . .

disaster itself is attributed to nature." Hewitt made clear that a crucial piece of the puzzle was missing and a fresh and politically engaged perspective from the social sciences was badly needed.[62] In 1994, Piers Blaikie, Terry Cannon, Ian Davis, and Ben Wisner—some of the best scholars in disaster risk reduction at the time—got together to try and settle the debate regarding the "naturalness" of disasters. They aimed to explain *why* disasters happen and wrote *At Risk: Natural Hazards, People's Vulnerability and Disasters*, still one of the most influential books in this field.[63]

The book title is a good summary of their central argument: natural hazards *plus* people's vulnerability *equals* disaster. It follows the logic that an earthquake in the desert, or a storm in the arctic, does not cause a disaster if nobody lives in these places. The authors claim that disasters occur when natural hazards meet *vulnerable* people. That generally refers to people who live in unsafe conditions, such as poverty and fragile houses. But, vulnerability, they contend, is not simply a present condition in people's lives.

In fact, unsafe conditions are typically the result of dynamic pressures. These dynamic pressures include lack of local resources, such as insurance or a decent income. They also include some macro conditions, such as unemployment, insufficient welfare and social networks, lax regulation, and inappropriate planning. Yet the authors do not stop here in their cause-and-effect reasoning. Instead, they assert that these dynamic pressures are rooted in social and political contexts. Danger, thus, can be traced down to root causes. At the heart of human vulnerability, they argue, is social inequality, marginalization, exclusion, discrimination, neglect, and other forms of social injustice.

As a matter of fact, many of the approaches developed in Latin America, in continuity with the vulnerability theory, benefitted from a Marxist or neo-Marxist perspective that positions socioeconomic inequalities as a root cause of vulnerability.[64] "The neo-Marxist perspective," writes Mark Pelling, "was to view disasters as deeply embedded within the social structures that [researchers] believed shaped everyday development experience."[65]

The general approach was easy to prove. Sufficient studies have found, for instance, that disasters hit the poor hardest.[66] For those who have insurance, financial capital, mobility, education, political influence, and other resources, a natural hazard is simply an inconvenience.[67] For the have-nots, the same event has dire consequences.[68] As a 2010 headline in the French newspaper *Le Monde* concluded: "In the face of nature's rage, humans are not all equal."[69]

FROM DIVINE PUNISHMENT TO DISASTER CREATION

Blaikie, Cannon, Davis, and Wisner argue that natural hazards simply trigger destruction. Nature lights the fire, while social, economic, and political pressures kindle its flames. The authors call their explanation of disasters the pressure-and-release (PAR) model. From this perspective, disasters are not the *cause* of disruptions in prevailing conditions but rather the *result* of the status quo. According to their model, vulnerability progresses through time until it meets the hazard. When the hazard collides with vulnerabilities, *boom*! Disaster ensues.

This proves to be a powerful insight. It means that disasters are not natural but created by humans. How? Through poor oversight and decision making, negligence, recklessness, and other preventable errors, such as building in dangerous areas and failing to provide healthcare, education, and insurance to marginalized groups.

Consider poverty in Central America. When, in 2001, two earthquakes struck El Salvador within thirty days, the poorest residents suffered the worst consequences. Of the six hundred people killed by landslides in Santa Tecla, and the hundreds of people in La Paz who lost their homes, most lived in slums and informal settlements.[70] Similarly, when Hurricane Mitch thundered through Central America in October 1998, the worst effects were suffered by low-income residents located in informal settlements along riverbanks.[71] When an earthquake hit Guatemala in 1976, the slums were so much more severely affected than other areas, which led some scholars to call it a "class quake."[72]

Consider also the marginalization of social groups, such as indigenous populations. In many parts of the world, indigenous people live in poverty and have little to no political representation. Their communities lack sufficient access to well-paid jobs, quality education, services, and infrastructure. Not surprisingly, indigenous communities are frequent victims of disasters in Peru, Bolivia, China, Honduras, and many other countries.[73] As a 2018 study on the impact of disasters in Canada states, "Indigenous peoples affected by natural disasters suffer a double indignity: their lives most affected, their voices least heard."[74]

Finally, contemplate the issues of segregation and neglect. In 2005, Americans had a vivid reminder of the proximity between racial segregation and disasters. When Hurricane Katrina hit New Orleans, neighborhoods with the highest proportions of African Americans suffered the greatest consequences. Those who did not have the means to evacuate the city were trapped in the chaos that followed Katrina's passage. Of the almost 1,500 people who died in Louisiana, 56 percent of them were black and 64 percent were over sixty-five.[75] Blacks and Latinos in the United States were also disproportionally affected by the COVID-19 pandemic that began in 2020.[76] Having insufficient access to healthcare, they often found themselves with fewer resources to be tested, supported, and treated.

Americans have also been reminded of the proximity between disasters and neglect. After Hurricane Harvey hit Texas in 2017, the network of drainage channels and reservoirs collapsed. The city then blamed the county for the security oversights, while the county blamed the city. In the end, both accused Congress of failing to construct a third reservoir, dam, and dike for Galveston Bay.[77]

Blaikie, Cannon, Davis, and Wisner's idea of vulnerability is one of the best tools we have for reducing risks. It helps us avoid the emphasis that has often been placed on natural hazards as the cause of disasters. By looking at both terrestrial causes, and the effects of institutions and human action (or inaction) on disasters, the concept of vulnerability also enables us to reduce the influence of superstition on policy and decision making. It helps us avoid attributing disasters to supernatural powers, and it emphasizes the accountability of those in positions of authority by highlighting their oversights, poor decisions, and negligence. Most importantly, the PAR model helps us identify the type of social injustices that lead to disasters.

However—as we shall see in the last section of this chapter—the idea of vulnerability also raises significant questions that are difficult to answer.

WE ARE RESPONSIBLE FOR DISASTERS, BUT WHO IS *WE*?

The notion of vulnerability implies that societies are responsible for disasters. Ultimately, it is the economic, social, and political system at a given place and time that puts people at risk. For instance, black and Latino Americans who have historically been denied quality education, access to well-paid jobs, healthcare, and other resources, often pay the highest price when disasters strike the United States.[78] As we saw before, the same applies to rural residents, slum-dwellers, indigenous communities, and various social groups in other countries.

But what happens when a disaster is caused by effects created by (or in) "other" countries or societies? Can we, for example, hold the Cuban authorities responsible for the climate changes that are intensifying the tropical storms on their island?

I don't think we can. The Global Carbon Atlas places Cuba seventy-first on the list of most polluting countries (calculated as per-capita CO_2 emissions originating within a country's territory).[79] From this vantage point, people like Mr. Cruz are not so much victims of a local system as they are victims of the international politics that have led to a commercial embargo in Cuba. Besides, Cubans, in this way, are also victims of the polluting cars, airplanes, and machines used in China, Europe, and North America that produce carbon emissions that contribute to global warming.

Climate change is not the only threat that travels across borders. Water and air pollution, deforestation, viruses, and overexploitation of resources have increasingly transnational consequences. Approximately 175,000 square kilometers of forests were cleared in the Brazilian Amazon region between 1990 and 1999.[80] This deforestation will not only affect Brazilians but also every other human being—and generations to come.

In that same way, wars spur mass migration and affect countries that have little to do with the sources of the violence. Since the Syrian civil war started in 2011, over one million Syrians have migrated to neighboring Lebanon.[81] Meanwhile, Venezuela's economic and political crisis, which began in 2016, has led almost six hundred thousand Venezuelans, many of whom claim refugee status, to migrate to Colombia.[82]

It's hard to hold local and national authorities responsible for threats that loom from beyond their jurisdictions. To explain why several disasters happen at once, we must adopt a very broad consideration of cause-effect relationships. Let me be clear: this does not mean we should absolve local leaders of responsibility. But as in the cases of Cuba, Lebanon, Colombia, and the Central American caravan, local leaders often share the responsibility for current threats with foreign warlords, businesspeople, and politicians.

There is no doubt that politics and politicians play a fundamental role in the creation of vulnerabilities. They also have a strong influence on the way disasters are defined.

FROM ACTS OF GOD TO ACTS OF LAW

On February 27, 2010, a mighty earthquake measuring 8.8 on the moment magnitude scale struck Chile. It was one of the strongest seismic activities reported in recent history. A few hours later, a tsunami hit the Maule and Bío-Bío regions, as well as Robinson Crusoe Island. The earthquake and subsequent tsunami killed 521 people and damaged or destroyed about 650,000 homes.[83] A terrible national disaster, you might think. Or was it?

Answering this question requires a brief review of Chilean politics, as this earthquake and tsunami occurred during the transition between two presidents—Michelle Bachelet and Sebastián Piñera. Bachelet, a socialist elected in 2006 as part of the center-left coalition Concertación, would finish her first mandate eleven days after the quake. Piñera, the incoming president, was a right-wing billionaire and the first conservative president to be elected in Chile since Augusto Pinochet's dictatorship ended in 1990.

In 2016, my colleague Kevin Gould and other researchers meticulously traced the chain of decisions made after the earthquake and tsunami hit.[84] They found that within a few weeks, the Chilean government—caught in transition between Bachelet and Piñera—adopted three distinct interpretations of the events. These interpretations were used to justify three different responses.

With houses and infrastructure still wet from the tsunami, President Bachelet's administration scrambled to project a reassuring image, as though the situation was under control. At first, Bachelet avoided invoking special measures. Chileans remembered the frequent abuses of states of exception during the Pinochet dictatorship, and government officials claimed that their agencies could handle the situation. Initially, Bachelet wanted to avoid declaring a state of emergency and deploying the army. In this first interpretation of the earthquake and tsunami, the government proclaimed that the event was serious but not a disaster. It could be managed with ordinary government measures and resources.

Yet, experts and journalists soon revealed that government agencies had failed—not only to predict the tsunami and organize evacuations but also to recognize the full extent of the damage. In many cities, the tsunami was followed by vandalism, looting, and violence. As the state's incompetence became clear, public pressure mounted for radical measures.

The government was forced to accept a second interpretation: that the event was a disaster beyond its current capacity to handle. Bachelet then declared a state of catastrophe in several regions. In the affected cities, civil rights were suspended, and the military was called in to restore order. In Concepción, the country's second-largest city, people's freedom of movement was limited to six hours per day. Curfews were also imposed in other affected cities and enforced by the army.

Less than two weeks after the disaster, Bachelet finished her mandate and President Piñera took office. Since his inaugural speech, he projected the idea that his administration was in charge of reconstruction. His right-wing government adopted a more classical neoliberal approach and reframed the situation a third time: the "natural" disaster was too complex and costly for the state to manage, so reconstruction would be more efficient if left to the business sector and the market.

Within the span of a few weeks, the same event had been framed as a delicate yet manageable incident, a disaster requiring the state to expand its powers, and a catastrophe so beyond the state's capacity, that it could only be handled by more efficient market-based solutions.

This scenario illustrates a problem with the notion of vulnerability. The scientific approach to understanding disasters has led many scholars and decision makers to believe that the label "disaster" is objective: an event is either a disaster, or it isn't, depending on vulnerability conditions that can be measured and traced on time. But as the Chilean case shows, disasters do not simply happen; instead, they are *declared to have happened.*

To fully grasp this concept, we must remember that houses are destroyed and people are killed every day by natural events. The Food and Agriculture Organization of the United Nations estimates that small-scale tragedies caused by climate events accounted for more than half of human losses in Latin America and the Caribbean between 1990 and 2014.[85] Most of these events were not labeled "disasters" by organizations and governments. They were not reported by international media and did not give rise to special laws or provisions. Why not?

"Disaster" is a label adopted by people in power in the face of events. Often times, the use or avoidance of this—and other labels, such as "state of emergency"—is linked to political agendas. Thus, disasters are not objective occurrences but dynamic social constructs, shaped by social and political conditions. Much like "war" and "genocide," "disaster" is used or applied according to social norms and political circumstances. Disasters are not acts of God, as some religious leaders have suggested. More often than not, they are acts of law.

The way authorities explain tragic events affects how (and whether) victims are supported. It also shapes vulnerability reduction plans and emergency actions. But as we will now see, the manner in which authorities explain and communicate risk is equally important.

ONE HUNDRED YEARS OF CONFUSION

A few years ago, Pia Rinne and Anja Nygren, two researchers from the University of Helsinki, examined how flood risk was explained in Tabasco, Mexico. They found that different explanations of risk have prompted a variety of responses. In 1999, the dominant discourse on flood control prioritized technological procedures. But years later, stakeholders and the media focused on the benefits of outsourcing risk prevention measures to the private sector, on the premise that this would lead to further development.

After 2007, narratives primarily focused on urban planning, relocation of people from disaster-prone areas, social resilience, and adaptation. As Rinne and Nygren explain, there was a "push to make local residents more capable to live with water," which transferred the responsibility for flood management to these residents. This narrative also helped to legitimize certain political agendas, including "evictions of informal residents from settlements near the city centre . . . justified on the grounds that they were necessary to save the city from future disasters."[86]

Language has been a powerful tool in Tabasco and other places of the Global South to pursue planning and managerial agendas in the face of risk. How risk is explained also determines disaster politics in rich countries. In the 1960s, the U.S. government sought a tool to balance two potentially conflicting objectives: protecting American citizens from floods and avoiding overly stringent regulations on land use.[87] The National Flood Insurance Program needed a tool to communicate risk to citizens and map it in planning documents. In response, the government introduced the abstract concept of the 1 percent annual exceedance probability, or AEP flood, which eventually led to the emergence of other equally abstract terms, such as the "hundred-year flood" and "hundred-year floodplain." These notions were later adopted by authorities in Canada and other industrialized countries. Today, they're used to map risks in cities, plan disaster responses, calculate insurance premiums, and make other significant decisions in the face of risk.

One of the objectives of adopting these terms was to create awareness among citizens about the risk of floods. But in this regard, the tool has often backfired. Most people think "hundred-year floodplains" are areas that get flooded about every hundred years—which doesn't seem very risky.

This is false but difficult to correct in plain language. In an article meant to clarify these terms, the Canadian Broadcasting Corporation (CBC) described hundred-year floodplains as areas "likely to flood once a century."[88] The journalist's explanation was well-intended but misleading.

In reality, the AEP concept refers to the probability of a flood event happening. It does not mean that one hundred years are likely to pass between each flood. Here's how the AEP works: Experts examine data and estimate values corresponding to amounts of water in flooded areas each year. This way, they estimate the AEP for various flood magnitudes. So when they claim that a certain zone is a hundred-year floodplain, they mean there is a one in one hundred chance that next year's flood will equal or exceed the 1 percent AEP flood. A hundred-year floodplain, essentially, describes an area subject to a 1 percent probability of a certain size of flood occurring in any given year.[89]

The hundred-year flood designation is an estimate of the long-term average recurrence of the event,[90] but it does not mean that we will have one hundred years between each flood of this (or greater) magnitude. Consider the example given by the U.S. Department of the Interior and the U.S. Geological Survey: If experts had one thousand years of streamflow data, they would expect to see about ten floods of equal or greater magnitude than the hundred-year flood. Yet these floods would *not* occur at one hundred–year intervals. In one portion of the one thousand–year record, there could be fewer than fifteen years between "hundred-year floods," while other parts of the record show 150 or more years between them.[91]

Misunderstandings about the AEP and other abstract concepts often lead people to underestimate their chances of being affected by floods—with dire consequences. Let us assume that you obtain a thirty-year mortgage to buy a house in a hundred-year floodplain. Over the course of the mortgage, your house has a 26 percent chance of being inundated at least once.[92] Studies have found that this is higher than what thousands of Americans perceive the risk to be. In fact, there are now more than 40 million Americans living in hundred-year zones.[93]

There is another reason why urban plans that depict hundred-year floodplains have, in some cases, backfired as measures to create risk awareness among the public. Many people find statistics and probabilities about natural hazards suspiciously precise, raising doubts about their legitimacy and the possible agendas behind them. A recent study found that when residents in Texas were asked whether they agreed that scientists could accurately assess the size of a flood, several participants exclaimed, "Who do they think they are, God?" The researchers concluded that maps (like levees) can create a false sense of security: some people believe that, since there's no way of knowing when a flood will take place, it probably won't happen to *them*.[94]

AS IF THEY WERE A YEARLY SURPRISE

The Nobel laureate Gabriel García Márquez once wrote that when he was a child, the heat in his tropical town was "so implausible" that "the adults complained as if it were a daily surprise."[95] His childhood memory relates to a problem of risk perception that happens when we accept the idea of vulnerability. In the PAR model, disasters are explained as exceptional events that occur after unsafe conditions have accumulated—often over long periods of time. Most academics, disaster consultants, and humanitarian workers tend to treat disasters like rare detonations. In the PAR model, for instance, there is a moment *before* and a moment *after* the disaster: unsafe conditions (and vulnerabilities) accumulate and are released by a natural hazard triggering the disaster moment. But, in reality, disasters tend to repeat in the same places—and rather quickly. For this reason, every post-disaster moment is also a predisaster moment.

This is particularly important because it means that the progression of vulnerabilities is not linear. It is often punctuated by catastrophic events that leave institutions and societies even more fragile. This reminds us that humans' short-term memory is a crucial cause of disasters. Consider, for example, meteorological events. Seasonal monsoon winds have existed for millions of years. But much like García Márquez's countrymen, people in India, Pakistan, and other countries in Asia, complain about them "as if [they] were a daily surprise."[96] What should actually be surprising is the fact that the heavy rains brought by monsoon winds provoke seasonal damages, deaths, and injuries.[97] A similar phenomenon happens with the weather patterns known as El Niño and La Niña in the Pacific. They bring disturbances that repeatedly trigger droughts and floods in Central and South America. Yet Latin Americans tend to forget these effects. They continue to build cities and settlements in places that are frequently put to the test by natural events—and then resume complaining about disaster as if it were a daily surprise.

Actions taken after a natural hazard occurs are the same actions taken *before* another, similar event takes place. This might seem obvious, but as the American floodplains show, it is not clear to everyone. As a matter of fact, many people tend to underestimate the probability of extraordinary events happening again, with similar intensity—especially in the short and medium terms.

Part of the reason is linked to the way we have been educated about disasters. We are inclined to view them as events that cause great impact but have infrequent occurrence and low probability. This is a viewpoint that we must increasingly challenge in the face of climate change. Disasters repeat more and more in the same places—although millions of people react to them as if they were unexpected surprises.

Not only are natural hazards considered unpredictable, but the very causes of fragility that lead to disaster also take organizations and authorities by surprise.

A VERY MACHO RECONSTRUCTION, AND OTHER ROOT CAUSES OF VULNERABILITY

To mitigate the inefficient government response when earthquakes hit El Salvador in 2001, a number of charities stepped in. Several international organizations moved to El Salvador to build houses, schools, and community services. They soon encountered two problems: thousands of destroyed homes and a culture entrenched in machismo. Some charities seemed better prepared to deal with the former than the latter.

Scholars know that women in the region are particularly vulnerable to—and less likely to rebound after—disasters. They often lack formal jobs and are excluded from social welfare, which culminates in higher unemployment. They have less political influence and are less likely to have decision-making roles. Instead, women are more likely to assume additional family responsibilities, such as caring for children and the elderly. Women are also less likely to own property and have access to credit. This limits their flexibility to move and find better jobs or homes in other areas. Women in the region are also frequent victims of violence and assault. In the face of disasters, they fail to receive medical attention and protection against sexual and domestic violence.[98]

In recognizing these social disparities, many international agencies, multinational consultant offices, and charities have adopted gender-equality principles in reconstruction and development initiatives across Latin America and the Caribbean. For example, a project funded by the European chapter of the Red Cross—and conducted in El Salvador after the 2001 earthquakes—adopted this approach. It was initially presented as an ideal opportunity to create an exemplary community and model settlement in the region. But, as we shall see, social change does not always come without consequence.

Soon after the earthquake, Alicia Sliwinski, a professor at Wilfrid Laurier University, studied the impact of women's participation at the time of the Red Cross project in El Salvador. Sliwinski found that women had been invited to participate in housing construction activities. Female beneficiaries were also given home property titles and space to voice their needs. But social tensions soon emerged in the community as women took on different roles. Disputes between women and within couples emerged. Many felt that traditional family structures were being disrupted. Eventually, the tensions led to serious problems within and between households. At one point, one of the women reported to Sliwinski, "There is no community here, people are all hypocrites." This is presumably not the "model community" the Red Cross project hoped to create.[99]

Sliwinski's findings shed light on the tensions that emerge when rapidly trying to change the social conditions that lead to vulnerabilities. Neither I nor Sliwinski pretend that projects should perpetuate inequalities or unjust gender roles, of course—but Sliwinski's study does demonstrate that shifting ideologies and deep-rooted social practices is difficult. It is a challenge that should not be underestimated in postdisaster contexts, though international agencies often fall into this trap and approach reconstruction projects as opportunities to quickly redress social injustices.

There are many reasons why it is not easy to change socially embedded practices and ideologies. One of them is that the timelines typically used for postdisaster interventions are too short to produce the type of change needed for redressing social injustices that lie behind vulnerabilities.

The Salvadoran reconstruction case highlights another significant problem with the vulnerability approach and the PAR model. By linking unsafe conditions to deeply ingrained traits that range from social behavior to cultural and institutionalized practices, the approach might lead us to believe that there is little we can do in the short term to help or protect vulnerable people without producing radical changes that might misalign with their cultural traditions. In fact, patterns of segregation, exclusion, and discrimination are even less likely to change amid the angst that follows a disaster.

My point here is that recognizing the root causes of vulnerability is a key intellectual exercise. But it is one that does not easily translate into effective measures to reduce risk. Most Haitians and Cubans I have worked with, for instance, are justly aware that trade policy and politics dictated by neoliberal American leaders is a root cause of poverty in their territories.[100] But they also know that they do not have any influence in international politics, so recognizing root causes is of little use for them in reducing risk in their countries. In many cases, this recognition of the impact of international forces on disaster risk creation even translates into a sentiment of impotence and dependence.

Now, is history really at the heart of current vulnerabilities, as the PAR model suggests? In 2001, Gregory Bankoff, then a researcher at the University of Auckland, and a known critic of the vulnerability approach, argued that vulnerability is part of a Western discourse of otherness that Europeans have adopted while observing the Global South. "Central to the [vulnerability] perspective," writes Bankoff, "is the notion that history prefigures disasters, that populations are rendered powerless by particular social orders that, in turn, are often modified by the experience to make some people even more vulnerable in the future."[101]

If people's vulnerability is rooted in historic social conditions, ideologies, and institutionalized practices—as most vulnerability theorists imply—how can we rebuild *immediately* after disasters? Or, how can we reduce vulnerabilities *now*? These are particularly puzzling questions for architects, urban planners, and other professionals interested in disaster risk reduction.

When I explain the PAR model and vulnerability approach to my architecture and urban planning students, they feel rather impotent. They find that the projects and policies they are able to develop can hardly reduce vulnerabilities in the short and medium terms. I need to explain to them that vulnerabilities can also be seen as lack of resources (material and otherwise).[102] Material resources include housing, infrastructure, and services. Soft resources include education, decision-making capacity, political representation, education, information, insurance, and employment. Vulnerabilities, in this sense, can be seen as lack of "capabilities" (in Amartya Sen's strict use of the term);[103] that is, broadly speaking, people's freedom and capacity to choose and have access to hard and soft resources, and by extension, their freedom to pursue opportunities that add meaning and value to their lives.[104] Here, freedom and the capacity to choose are crucial components of people's agency.[105] Projects and policy can help increase access to these resources and professional skills. This reassures my students, but only to a certain extent. Seeing the world through the lens of vulnerabilities entrenched in history is useful for detecting social and political problems, yet ineffective for identifying the type of change that is needed in the short and medium terms.

A focus on vulnerabilities carries another problem: it sometimes leads to victimization of citizens. By stressing the influence of segregation, marginalization, poverty, and exclusion, the vulnerability approach leads decision makers to focus too much on what people or regions lack, underestimating the strengths that local residents possess. "Tropicality, development and vulnerability," writes Bankoff, "form part of one and the same essentializing and generalizing culture discourse that denigrates large regions of the world as disease-ridden, poverty-stricken and disaster-prone."[106] This view often paints people in the Global South as having little agency and capacity to change their own fate.[107]

Disaster researchers in Latin American countries have found, for instance, that the "vulnerabilization" of risk has led development agencies to treat women as passive victims of disasters. Some scholars, therefore, argue that there is a need for a new approach to risk based on people's agency rather than assumptions about their fragility.[108]

Following recent feminist approaches, some authors have even argued that vulnerability can also be seen as a strength and have reflected on "the power of vulnerability."[109] According to Jason von Mending, a professor at University of Florida, and Heidi Harmon, the mayor of San Luis Obispo City (California), real transformative change requires recognizing the value of vulnerability.[110] This narrative sustains that those in vulnerable positions have particular perspectives that enable them to identify solutions that others cannot.[111]

In the following chapters, we will see other examples of how an approach based on vulnerabilities has also served to justify Northern and Western interventions in disaster risk reduction, and by extension, the adoption of Northern concepts in the Global South. We will also observe how an overemphasis on vulnerabilities has led both local and international decision makers to adopt paternalistic methods that eventually limit people's capacity to decide for themselves.

A vulnerability-based explanation of disasters also raises a significant question: Are all victims of disasters *really* vulnerable?

DISASTERS DON'T CARE ABOUT WEALTH

In December 2017, Paris Hilton tweeted: "This wild fire in [Los Angeles] is terrifying ☹," adding, "My house is now being evacuated to get all of my pets out safely."[112]

Hilton tweeted her distress as firefighters were battling to save mansions and estates in wealthy BelAir and other parts of California, where celebrities and billionaires—including Rupert Murdoch, Ellen DeGeneres, Lionel Richie, and Elon Musk—own homes. And in November 2018, when massive fires broke out once again in California, the town of Paradise, home to twenty-six thousand residents, was reduced to ash.[113] More than seventy people died, and the evacuation of celebrities, pets, and other well-off residents was repeated. Reflecting on the recent wildfires that ravaged California, Governor Jerry Brown claimed in 2018, "This is not the new normal. This is the new abnormal."[114]

Abnormal events such as these wildfires challenge the very principles that help us explain disasters today. The vulnerability concept argues that disasters happen when people are vulnerable. It follows that those who are poor, excluded, and fragile are more likely to be affected by disasters. Unequal distribution of opportunities is at the core of Marxist and Neo-Marxist approaches to disaster risk reduction. But this principle does not easily help us explain why disasters also affect well-off people in California, Japan, Canada, Australia, and other developed countries. In fact, if you already bought the idea that disasters are caused by rooted vulnerabilities among citizens in Haiti, Peru, Vietnam, Honduras, El Salvador, Bangladesh, India, and other poor countries, you probably feel uneasy applying a Marxist view of vulnerability to the disaster that affected Ms. Hilton and her spoiled Chihuahuas.

Pelling writes that neo-Marxism in disaster studies is "criticized for over-privileging economic class in its analysis and for failing to identify the importance of individual agency in the production of vulnerability."[115] Perhaps you also find that you are overstretching the pressure-and-release concept when trying to explain the cause of the Tōhoku earthquake and tsunami that hit Japan in 2011, which killed almost sixteen thousand people (you probably saw the images of the expensive boats overflowing the levees built in the town of Miyako, near the Hei River).[116]

Reporting on the 2018 wildfires in California, a subheadline in the *New York Times* read "Wildfires don't care about wealth or status."[117]

Which of the headlines is more appropriate to describe disasters today: the one in the *New York Times*, or the one in *Le Monde* in 2010 ("In the face of nature's rage, humans are not all equal")? They certainly point in two different directions. Where *Le Monde*'s headline embraces the classical precepts of vulnerability, the *Times* seems determined to challenge them.

This contradiction keeps puzzling experts, who find that the arguments used to explain disasters in some places need to be overstretched when applied to other contexts. But in reality, vulnerabilities are not created through social injustices alone, as one might initially think. Exposure to hazards is not always the consequence of marginalization. Fragility and exposure are also created by humans' reckless actions toward the environment. And this recklessness is often found among the wealthiest and privileged. This is the case of residential developments in California, and more generally, suburban growth in North America. As a matter of fact, it is estimated that since 1990, 60 percent of new homes in California, Washington, and Oregon have been built in close proximity to forests and natural features.[118]

Of course, there is another form of human action that increasingly contributes to these fires: climate change. In a warming planet, there are longer seasons and drier areas where fires are prone to ignite.[119] It has been found that in U.S. western states, for instance, the average fire season is now eighty-four days longer than it was in the 1970s.[120]

Disasters in affluent neighborhoods are also happening elsewhere. In 2018, more than fifty people died in the suburbs of Athens, Greece, from a forest fire.[121] According to a study by Greenpeace, in the last twenty-four years, urbanization in the Spanish littoral has increased by 57 percent.[122] By moving near the coasts and forests, scores of Spaniards, Greeks, and Americans are putting themselves at risk—with the endorsement of authorities that provide them construction permits. Contemporary societies thus put some of its wealthiest members at risk, alongside the have-nots.

THE UNNATURAL DISASTER CAUSED IN NATURE

The theory proposed by Blaikie and his colleagues is often seen as having a proper balance between human and physical systems.[123] But the idea of vulnerability still raises another question: Are disasters only those events that affect people?

Most vulnerability theorists consider that a disaster occurs "when a significant number of vulnerable people experience a hazard and suffer severe damage and/or disruption of their livelihood system."[124] But isn't the destruction of ecosystems, as we know them today, a disaster on its own—regardless of people being directly affected or not? If we adopt the idea of vulnerability to explain disasters, it is difficult to see the melting of artic ice, for example, as a disaster trigger. Its impact on people is indirect (mostly through sea level rise) and is mediated by several contextual characteristics, such as the dependence of people on one source of income or the proximity of homes to the shore.

Yet several ecologists and environmentalists have no problem accepting that ice melting, coral reefs disintegrating, and the effects of climate change are all disasters on their own—whether they affect people or not. In 2020, some experts calculated that fires in Australia killed about twenty-five thousand koalas. Ten thousand feral camels were expected to be shot and killed (some estimate that a total of one billion animals perished across Australia during the fires).[125] Just as ecologists and environmentalists have focused on the hazards' effects on nature, sociologists, anthropologists, and other social scientists have focused on their effects on humans. But should we perpetuate the separation between disasters that affect people and those that disrupt ecosystems?[126] What would the story of a disaster be if told not by humans but by koalas and animals at risk of extinction due to our endless pollution? I suspect that people's vulnerability would not be as important as it is for us.

Contemporary cities and human settlements are physical representations of the ways societies treat their more fragile members. But they also emerge as the results of the ways societies and institutions let the most privileged use their power. These privileges are increasingly responsible for the deterioration of ecosystems and the atmosphere. From this perspective, an urgent need to protect people and the environment is "the new abnormal." In this book, I argue that they are no longer two different things.

In the following chapters, we will see how disasters and risks can serve various purposes and advance all types of agendas. This includes the laissez-faire attitude of political leaders in Haiti toward the proliferation of informal settlements, as well as the delegation of responsibilities—assumed by governments—to the private sector in the UK and Latin American countries. We will see how risks, climate change, and disasters can simply become opportunities to tax more, grab land, expropriate the most vulnerable, help partisan groups, and—through the all-too-common declarations of states in emergency—advance policies entrenched in ideology. In the next chapter, we will see that many of these actions are easy to implement when they are called "modern."

2 Change

"They Want to Build Something Modern Here"

There are also impersonal drawings, of undefined purpose: One drawing will decree that a bird be released from a tower roof; Another, that a grain of sand be withdrawn (or added) to the innumerable grains on a beach. The consequences, sometimes, are terrifying.

—Jorge Luis Borges, "The Lottery in Babylon"

DISASTERS CHANGE EVERYTHING

Musi-café had become a popular place. Locals gathered to drink coffee during the day and beer and spirits in the evening. Most people in the town of Lac-Mégantic knew the café owner, the servers, and the musicians who played at night. The main road, a short walk from the lake, was filled with shops. Like almost all cities and villages in the French-Canadian province of Québec, Lac-Mégantic's Catholic church was surrounded by the best houses in town. Railroads crossed the central neighborhoods, which had a historic character that made residents proud. By and large, Lac-Mégantic was a quiet place.

But on July 6, 2013, a seventy-four-car freight train carrying oil derailed in the city's downtown core, triggering an enormous urban fire. Violent flames killed forty-seven people, orphaned twenty-one children, and destroyed the city center. The intense and sudden blaze destroyed more than thirty buildings, and almost forty more had to be demolished later due to oil pollution. This was one of the deadliest rail accidents and urban disasters in Canadian history. More than half the victims had been enjoying an evening at Musi-café.[1]

For Méganticois, life changed forever. Yet, the chain of events that followed is all too common in postdisaster situations.

First, residents and journalists sought explanations. Experts appeared on the news and deplored train security standards (we eventually learned that security oversights were at play). Then, outraged citizens and authorities embarked on a search for *who* was responsible for the disaster (most fingers were pointed at the rail company and rail transportation agencies).[2] Soon, political leaders visited the area, in rolled sleeves and khaki vests, to condemn the safety oversights.[3] A few days later, local and national journalists reported the "miserable conditions" victims found themselves in. Fuelled by these images, charities collected money for them. Artists and celebrities promised donations, while the government vowed to secure generous funds. Authorities took the stage to proclaim that "rapid sustainable reconstruction" would follow, along with variations on the standard we-must-guarantee-this-never-happens-again.[4] Local leaders hired designers to draft sustainable development plans. Appointed and voluntary designers, engineers, and inventors came out with innovative solutions to "rapidly" rebuild infrastructure and buildings. Unfinished (but always uplifting) plans and 3D renderings were presented to the public. The city even set up tax credits to incentivize the construction of green-certified buildings, including those bearing endorsements like LEED (see chapter 3) and Novoclimat (a local energy-efficiency certification).

Months later, residents complained that the generous donations hadn't reached them. Trains full of oil continued crossing cities in the region. Journalists moved on to other urgent matters—only to show up again on the first anniversary of the disaster, and then again on the fifth. Sustainable development plans and drawings were shelved. Charities left. Benefit concerts and other fundraising events became less frequent. Political attention shifted to other pressing issues (in this case, floods). Several years later, a few ribbons were cut—a green building here, a new facility there, and a piece of infrastructure nearby. In time, the public aspects of the disaster simply became additional material for historians (and the occasional scholar of unnatural disasters) to record.

For many, the whole thing had become a spectacle.[5] "All of the sudden, the issue was no longer political," observed a group of researchers. "The songs of goodwill became more important. In this way, those in power avoided criticism."[6] For Gilbert Liette, a professor at York University, the postdisaster process was "about emotionalizing resilience as a way to discredit contestation and instill acceptance and support for reconstruction efforts."[7]

After the Lac-Mégantic tragedy, senior citizens were particularly affected.[8] For many of them, years of history, culture, peace, and tradition were lost in a single night.[9] In the aftermath, it only took one sentence and one drawing to bar people from trying to recover them.

Soon after the Lac-Mégantic disaster, architect Pierre Thibault—one of the most celebrated in Québec—was commissioned to draft a town-planning document for a disaster-ravaged central district. His urban planning proposal was filled with all the typical sustainable development rhetoric. Objectives included preventing urban sprawl and promoting urban coherence, mixed use, space appropriation, and a "collective experience at the human scale." Thibault's drawings had all the features that would impress defenders of sustainability and resilience: urban alleys planted with wild vegetation; hidden parking spaces; brick, stone, and the occasional wood façades; driveways on grass pavers; smiling, good-looking figures riding bicycles; and of course, lots of trees, so leafy and numerous that they were often left translucent—like green ghosts—so as not to interfere with the drawings' main purpose: to display the architect's talent.

Thibault's plan was decidedly green and precise. There were provisions for the design of signs, fences, public furniture, and urban lighting. But something else was also prescribed in a rather unusual fashion. In one of

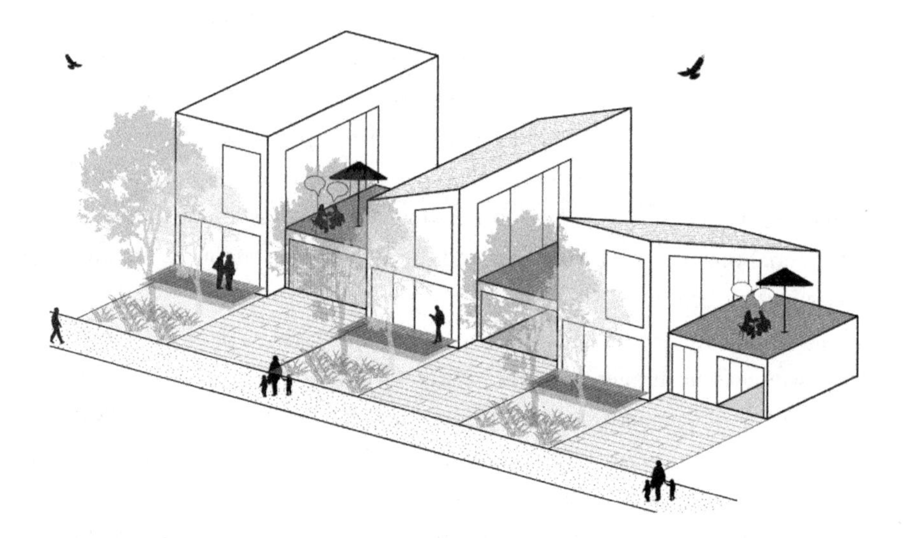

FIGURE 2.1 Architectural rendering proposed by Canadian architect Pierre Thibault for houses to be rebuilt in Lac-Mégantic.

Source: Pierre Thibault and Ville de Lac-Mégantic. *Annexe 2—Objectifs Et Critères P.I.I.A.—010—Secteur Résidentiel Du Centre-Ville* (Lac-Mégantic: Ville de Lac-Mégantic, 2015).

the passages on the specifications for residential buildings, Thibault wrote, "The architectural typology must be contemporary. Styles based on Victorian, country-style, rustic-style, colonial-style, or manor-style references are not permitted in this urban development. Any form of pastiche is prohibited."[10]

Figure 2.1 shows one of Thibault's drawings: his vision of what a contemporary house should look like. Arguably, this drawing serves less to propose a new house design than to prevent imitation historic houses from ever being built in Lac-Mégantic.

Fake neoclassical buildings are popular in Canada, yet the phrase in Thibault's plan passed almost unnoticed. Contemporary architects confer value on the idea of authenticity. Replicating the style of ancient and classic buildings is considered sacrilegious in mainstream architecture today—a form of fraud, even. Buildings and neighborhoods must have a contemporary look and be loyal to the aesthetics of their own time. Nonetheless, it is hard not to feel a sense of unease about Thibault's decision. Why?

To be clear, when destruction occurs, some form of change is needed—the suffering that follows is tangible proof. But authorities and affected residents often want different types of change. For instance, many of the senior residents I have spoken with in disaster-affected areas, whether in Canada, India, Colombia, Cuba, Haiti, or elsewhere, tell me that all they want is to have "things the way they were before." They wish they could to return to the lives, houses, and neighborhoods they inhabited *before* the disaster. In fact, several studies show that citizens who have spent years in a place that is later affected by disaster find comfort in recuperating heritage and recovering familiar spaces and references (whereas young people and recent arrivals are sometimes more inclined to seize the opportunity for radical change).[11] Disaster victims often value tradition and the culturally embedded rituals that constitute place attachment.[12] "Most people who survive a devastating disaster want the opposite of a clean slate," writes Naomi Klein in *The Shock Doctrine*. "They want to salvage whatever they can and begin repairing what was not destroyed."[13]

The intent to prevent citizens from rebuilding a house in, say, a Victorian-style pastiche carries significant moral weight. Of course, recreating the past is a contentious objective and one that is hard to achieve. But in Thibault's planned neighborhood it will be impossible. In Lac-Mégantic's new town center, a resident will not be able to build a new house that reminds her of the old-fashioned one she lost in the disaster.

For many years after the tragedy, Méganticois continued to struggle with provincial and federal authorities. To this day, they still ask authorities to modify the tracing of train lines and to increase security measures in rail transportation.[14] They keep pushing the government to adopt stronger measures to protect the heritage of their territory and buildings. As for Musi-café, it was rebuilt a few years after the disaster—in a contemporary look.

The Lac Mégantic story illustrates a common pattern that has been found in previous studies. Even when decision makers have enough power and resources to generate change, they may produce solutions that do not respond to the needs and expectations of affected citizens and communities.[15] They also underestimate the value that residents attribute to their history and traditions.[16] As in Borges's imagined city of Babylon, the consequences are sometimes terrifying.

Behind these efforts are politicians eager to see the type of development that generates votes and produces economic growth. Yet, the type of sustainable development authorities want often produces losses that residents resent. As the following case in Colombia shows, there is always a price to pay for being green.

"THE MOST SUSTAINABLE SOLUTION THAT BOGOTÁ HAS SEEN IN RECENT YEARS"

When I was a child, my grandparents took me on the pilgrimage expected of every Bogotano of their generation—to see the Salto del Tequendama, a 132-meter waterfall on the Bogotá River. The river crosses the mountain plateau atop which Bogotá sits (at 2.6 km above sea level). I remember two things about that trip. One is that the waterfall was much less impressive than the image I held in my mind (probably influenced by my grandparents' recollections of family picnics on the river's shore). The other is that the river smelled horribly of boiled eggs.

Nowadays, few adults would voluntarily take their families to visit the Tequendama Falls. For decades, the river's flow has been significantly reduced by the water needs of thousands of businesses and millions of Colombians. Furthermore, the river has become so polluted that the Salto del Tequendama is suspected to be one of the largest wastewater falls in the world.[17]

Many Colombians now fear that the river's fate will only worsen. At the heart of this fear lie the planning decisions made by Enrique Peñalosa, an intellectual-turned-politician who served two terms as mayor of Bogotá.

Toward the end of his first mandate, from 1998 to 2001, Peñalosa was a popular mayor. In his second mandate, from 2016 to 2020, things took a turn. Peñalosa was elected again, this time replacing Gustavo Petro, a leftist political leader and controversial former senator. In what is still a very conservative country, Petro had endured four years of tumultuous partisan rivalry and political deadlock. One of the casualties in these political wars was the new Plan de ordenamiento territorial (or POT), which all municipalities in Colombia were required to produce.

As it turned out, Petro's POT had not been officially approved when Peñalosa took office again in 2016. Due to the lack of a legitimate city plan, housing construction had been virtually frozen for several years—and given Bogotá's rapid demographic growth, this paralysis reached a critical point. Housing prices skyrocketed in the capital,[18] and with a lack of available land in Bogotá, subsidized residential developments moved to Soacha, a neighboring city closer to the Tequendama Falls.[19] This shift, however, only exacerbated traffic, pollution, and the commuting troubles of thousands of low-income people. Peñalosa needed to address the problem, yet a popular solution proved to be out of reach.

As an internationally recognized expert in sustainability, Peñalosa had been reelected because of his perceived capacity to deliver a trendy new urban model. Voters saw Peñalosa as an ecologist and a champion of sustainable urban practices. For many, he was the real architect behind TransMilenio, the rapid-transit bus system that heralded a modern age for Bogotá. In a city where traffic is unbearable, TransMilenio's achievements from 1998 to 2001 put Colombian public transit on the smart-city map. But, in his second mandate, a lack of urban land for residential development would become Peñalosa's greatest challenge.

After his first few weeks in office in 2016, and in response to mounting pressure for a housing solution, Peñalosa shocked constituents by announcing his plans to build new neighborhoods on the Van der Hammen reserve. The reserve holds a system of wetlands that host numerous species of birds, frogs, and butterflies native to the area. This ecosystem at the city's outskirts works like a sponge that regulates the Bogotá River's water flow.[20] Peñalosa's plan included building roads, housing, and services on several parts of the 1.4 hectares of protected land.

Ecologists were outraged.[21] Thousands of voters felt betrayed. Journalists, politicians, and experts denounced the plan's ecological cost. Few technocrats supported the idea of urbanizing a natural reserve that spans from the eastern mountains surrounding the city to the Bogotá River and its fragile hydrological system.

In 2018, Peñalosa was also at the heart of a controversial decision to cut down hundreds of trees in Bogotá, mostly in low-income neighborhoods. His administration claimed that the trees were sick and were not native to the area, but ecologists argued that they were cut down to make way for infrastructure projects. Now, many believe that Peñalosa, the sustainable development champion of the 1998–2001 period, is not so green after all.

The Van der Hammen area was flooded as recently as 2011. Several houses and hectares of crops were lost in that event,[22] and in the aftermath of the 2011 floods, it was revealed that a prestigious private university had built part of its campus on wetlands that should have remained untouched.[23] Environmentalists were further outraged when Peñalosa allegedly called the Van der Hammen area a *potrero* (Colombian Spanish for an insignificant urban fringe).[24]

Peñalosa insisted on moving forward with his plans. In 2019, he confirmed that he would modify urban regulations to permit construction of the wetlands. By then, he had come up with additional arguments to justify his plan. For instance, he estimated that constructing housing on the reserve would prevent 1.8 million tons of greenhouse gas emissions each year by reducing residents' commuting time to neighboring cities. His administration called the Van der Hammen project "one of the most sustainable solutions that Bogotá has seen in recent years."[25]

Construction in the Van der Hammen was expected to start in 2020, but elections were held in 2019. Millions of voters seemed to have remembered the elders' stories about the Salto del Tequendama, and so, the candidate supported by Peñalosa lost the elections. Claudia Lopez, the new mayor, is now busy with the COVID-19 crisis and is less enthusiastic about Peñalosa's plan.[26]

It's true that land is scarce in Bogotá, as it is in most major cities today. It is also true that urban development beyond city perimeters increases ecological footprints and commuting times. Urban sprawl makes inefficient transportation systems, services, and infrastructure. It contributes to pollution and deforestation, and as such, it accelerates global warming. Lack of available land for residential development pushes rents up, which affects the poorest and more vulnerable citizens most. Authorities in many other cities now face Peñalosa's dilemma. The decision is difficult: to restrict urban development and protect green belts and reserves—accepting the drawbacks of distant suburban development—or to facilitate construction of vacant land and green areas, prioritizing the supply of affordable housing closer to infrastructure and services.

This dilemma has several nuances in different cities and contexts, and it is influenced by numerous economic and political factors. But it is not absurd to frame the question in bolder terms: If you were in charge, would you help the poor by deregulating land use, or would you protect the environment at any price?

Regardless of your answer to this question, the Bogotá River case reminds us how difficult it is sometimes to reconcile environmental protection and social justice. It also highlights that ambitious urban schemes carry substantial consequences in both domains. Of course, this does not prevent authorities from deploying them. Instead, they often resort to well-known tricks.

FROM SHOCK THERAPY TO DEVELOPMENTALISM AND PATERNALISM

Ever since Michel Foucault wrote about the state's power in the 1970s,[27] we have been constantly reminded that authorities use fear to master plan space and society in such a way that controls their citizens. In 1998, we learned how this is done in cities and territories. That year, James Scott, a professor at Yale University, published a book that changed the way we see the impact of authorities in human settlements. In *Seeing Like a State: How Certain Schemes to Improve the Human Condition Have Failed*, Scott argues that when authorities deploy their master-planned schemes, they prioritize legibility, simplification, and standardization. Legibility implies the erasure of local practices, knowledge, and rituals. "A thoroughly legible society," Scott explains, "eliminates local monopolies of information and creates a kind of national transparency through the uniformity of codes, identities, statistics, regulations, and measures." "If we imagine a state that has no reliable means of enumerating and locating its population, gauging its wealth, and mapping its land, resources, and settlements," he adds, "we are imagining a state whose interventions in that society are necessarily crude."[28] This process also necessitates the invention, adoption, and promotion of units of measurement that are visible and can be standardized. Only then can citizens, villages, houses, trees, and other units be identified, observed, counted, recorded, aggregated, and monitored, argues Scott. Simplification of measures and standards requires the creation of a uniform, homogeneous citizen. To exploit resources, the state has to be able to map nature and settlements, as well as render them legible. Rubber-stamped standardization of housing solutions and urban features is a mighty tool in this process.[29]

Many other stories of domination via development have been reported by contemporary scholars. In *The Divide: A Brief Guide to Global Inequality and Its Solutions*, anthropologist Jason Hickel finds connections between different notions of development and forms of control in postcolonial times.[30] Throughout this book I contend that one of these notions, neoliberal orthodoxy, augmented the dependency of poor countries and put them at a disadvantage in the global marketplace (see the Salvadorian example outlined in chapter 1). It eventually led to inequalities between rich and poor countries.[31] At the same time, however, Hickel finds that development policies implemented after the end of colonial rule in many poor countries lead to "developmentalism." This approach focused on increasing the intervention of the state and reinforcing and protecting domestic markets through tariffs on imported goods. Developmentalism became in some cases an alternative to oppressive neoliberal policies promoted by the Washington consensus. But Hickel contends that, "developmentalism was not without its flaws." He argues that "modernisation came with significant costs." In many instances, "the focus on rapid economic growth sped up the process of commodifying human life and nature that had begun under colonialism."[32] Following patterns of domination adopted in colonial rule, traditional values and lifestyles were treated as a barrier to economic growth and social progress and were often purposefully eradicated.

Now, authorities' schemes in the face of risk are not always as callous. Standardized regulations in construction are often motivated by intentions to improve people's wellbeing and protect them from hazards. Floods, fires, earthquakes, and major public health incidents are seen as opportunities to update laws and regulations—most of which make us all safer.[33] After the September 11 attacks in the United States, for example, the National Institute of Standards and Technology made recommendations for construction code changes. They sought to increase safety in the event of buildings being hit by flying objects—a risk that had hitherto been largely ignored.[34]

Through regulations and planning, politicians and technocrats seek to control people's behavior to enforce order. But authorities also routinely regulate people's behavior to prevent potential self-inflicted harm and instead encourage self-enhancing actions—even when there are no apparent benefits or disadvantages for others.[35] Paternalism is another aspect of authorities' attitudes towards radical change in times of risk. The Lac-Mégantic case exemplifies how decision makers sometimes consider themselves to have greater wisdom about what is harmful or helpful to vulnerable individuals than the individuals themselves. We will see in following chapters that there are numerous scenarios where sustainable development initiatives (some deployed for paternalistic reasons) fail to respond to the real needs and expectations of the poor and marginalized.

In a world of unnatural disasters, political authorities are not the only ones resorting to legibility, simplification, standardization, and paternalism. In some cases, charities and aid agencies are behind similar schemes, equally convinced they have greater knowledge of what vulnerable people need than the people themselves. Let us now see what change means in humanitarian aid.

SUSTAINABLE DEVELOPMENT AND AID

When Hurricane Mitch swept Central America in 1998, significant parts of Honduras were flooded. An estimated seven thousand Hondurans died and thousands more went missing due to the destruction caused by torrential rain and wind.[36] More than 220,000 houses and several bridges were destroyed. Choluteca, a remote city in the south of the country, was partially under water. A few months later, Choluteca was inundated again—not by water this time, but by charities.[37]

More than twenty-four NGOs flocked to Choluteca, including Caritas, Atlas Logistique, Doctors Without Borders, CECI, and Iglesia de Cristo. Almost a dozen Christian organizations, mostly from the United States, were soon building houses (and, of course, churches) in Choluteca. International agencies and consultants also showed up, including the International Organization for Migration, UNICEF, and the Spanish and

Canadian Cooperation Agencies. Soon, a new neighborhood called Nueva Choluteca ("New Choluteca") was under construction on the outskirts of the old city.[38]

Charities built over two thousand houses in Nueva Choluteca. They didn't put much infrastructure or vegetation in place, though. They set up only about one hundred latrines, one temporary water tank, and a few roads that were eventually left unpaved. Space was allocated for a park, but no one bothered to plant trees or build a proper sports field. The NGOs seemed to be more focused on building churches and putting a roof over their worshippers' heads than creating the conditions for residents to find or sustain jobs, and for Honduran children to grow up healthy and well-educated.

The houses were very small. They also lacked washrooms and were delivered unfinished. Beneficiaries had to work in construction, so a series of construction brigades were organized by the charities. After the guided self-help was finished, residents were expected to complete their own houses—to paint them, finish the kitchens, and install doors, floor finishes, and other details. In a matter of two or three years, most residents upgraded their homes. They also built fences and planted trees.

In 2005, I travelled to Nueva Choluteca as part of a study on the performance of postdisaster reconstruction initiatives. When I walked through the place, there were no paved roads, clinics, parks, sewage treatment plants, public transportation services, electricity or water systems. Violent crimes were commonplace. Public health was a pressing concern. Some residents had even abandoned their "free" houses. Almost half them were jobless—all were poor.[39]

Many of the residents who abandoned the new homes returned to their original neighborhoods, precisely where floods had affected them in the first place. Some dismantled the new houses and brought the doors, windows, and roof tiles with them to reconstruct their original homes in old Choluteca. In 2017, the central government in Tegucigalpa was still promising to pave the roads in Nueva Choluteca.

Unfortunately, the failed charities in Honduras and the urban fiasco in Nueva Choluteca are not rare examples. I have found abandoned postdisaster houses built with donors' money in San Salvador (El Salvador), Cape

Town (South Africa), and Armenia (Colombia). In some cases they were given to disaster victims for free—well, almost free. Most charities required beneficiaries' labor. And beneficiaries don't just abandon houses. Water wells, temporary shelters, schools, and infrastructure projects constructed in the postdisaster rush, often under the premise of "sustainable development," are usually deserted as well.[40]

Back in 1987, the Brundtland Report had said that more aid is required to achieve sustainable development in poor countries. It had also prescribed its use: "Part of the increased aid should go directly to community groups, using intermediaries such as national or international NGOs."[41] Sustainable solutions to the mounting crises in the underdeveloped world were to be achieved with the help of charities. Why? Because they "are generally more successful at reaching the poorest," the report stated.

The enthusiasm for international NGOs helping "the poorest" is hardly surprising. Few things attract more donations than a major tragedy. It is estimated that following the 2010 disaster in Haiti, the Red Cross raised more than $450 million in the United States and more than $1 billion worldwide (all figures in this chapter in USD).[42] The Red Cross alone received more than $5 million for interventions in Lac-Mégantic.[43] A World Bank report estimates that as much as $4 billion was gathered for reconstruction after Hurricane Mitch.[44] For urban consultants and the aid industry, disasters are a key source of jobs, contracts, and organizational growth. There seems to be no better way to collect donations than a good disaster, covered by international media and amplified by the visit of a few celebrities.

International aid is commonly seen as a crucial component of achieving sustainable development in poor countries. Bilateral humanitarian assistance grew from almost $600 million in 1980 to almost $2.8 billion in 1998.[45] The World Bank estimates that net official development assistance and aid increased from about $42 billion in 1987 (when the Brundtland Report was published) to $162 billion in 2017.[46] During that time, the "official development assistance" spending by OECD countries followed a similar trend. But as the Choluteca case shows, and the following section explains, the effect of this aid is almost always deceiving.

DOES AID (ACTUALLY) AID IN ACHIEVING SUSTAINABLE DEVELOPMENT?

For the past few decades, intellectuals and politicians alike have debated the efficiency and pertinence of aid. Devoting part of a national budget to international aid is frequently criticized in industrialized countries' parliaments, as well as in international forums and academic conferences. But it is fair to say that—at least in commercial books—the debate has been largely animated by white, American, Ivy-League economists with a neoliberal orientation. Thus, it is a discussion that has mostly revolved around aid's capacity to foster economic growth and the establishment of free markets in poor nations—shellacked with a thin coat of "sustainable development" rhetoric.

The Ivy League controversy around the effectiveness of international aid for development reached its highest point in 2006, when two widely known scholars in different parts of Manhattan engaged in an intellectual debate fuelled by their respective bestsellers.

In one corner of this debate was Jeffrey Sachs, then a professor at Columbia University. His work was followed by Bono, Angelina Jolie, and other celebrities. Sachs had recently published *The End of Poverty: How We Can Make It Happen in Our Lifetime.*[47] With the help of a few superstars, he campaigned for increases in the percentage of budgets allocated for international aid. He had examples at hand: in 2006, for every $100 in national income, Americans donated just 18 cents in official development assistance to poor countries. The Swedish were donating five times as much.[48] Sachs argues that, with more money, he and his dream team could implement real solutions. Besides, research could be devoted to monitoring change and finding the best methods for success. According to Sachs, the poor are often stuck in what economists call "poverty traps." In other words, they are poor because they are poor. People living in slums, for instance, have difficulty escaping poverty because they pay proportionally more for services and basic goods than wealthier citizens. Market behaviors are typically not enough to lift them out of poverty. Foreign aid is necessary to break this vicious cycle and replace it with a virtuous one that makes residents in poor countries more productive.

Sachs also notes that "environmental degradation at local, regional, and planetary scales threatens the long-term sustainability of all our social gains." He makes a call to strengthen the World Bank, the United Nations, and other international agencies. And he invites us to embrace the UN-supported Millennium Development Goals. Sachs argues that as "we invest in ending extreme poverty, we must face the ongoing challenge of investing in the global sustainability of the worlds' ecosystems."[49] That's a lot of global, worldly, international rhetoric. But that is a view shared by many environmentalists and sustainable development activists today.

In fact, most defenders of aid (you may want to place representatives of the United Nations and charities here) claim that lack of economic growth prevents sustainable forms of development in poor nations. With aid, they say, we can prevent famines, public health problems, and environmental degradation. Aid can also help in reducing housing and infrastructure deficits. It can be used to improve education and reinforce governance in poor nations.

A patronizing argument from the defenders of aid is that if (or when) poor countries become wealthier, they will (finally) become environmentally conscious. People in poor countries, they argue, will start recycling and campaigning for less smog in cities once they have "properly developed." With some aid, they will "follow the example" of rich countries today. This narrative, linking environmental consciousness and economic growth, is popular among white American and European economists. But no one has expressed it in a more condescending manner than Steven Pinker, a Harvard professor and theorist of almost everything (from cognition to the notion of world progress).

Pinker is not necessarily a staunch supporter of sustainable development itself, but he seems to be interested in the environmental benefits of development. In *Enlightenment Now*, Pinker finds that "as countries first develop, they prioritize growth over environmental purity. But as they become richer, their thoughts turn to the environment." One of his most absurd examples: "If people can afford electricity only at the cost of some smog, they'll live with the smog, but when they can afford both electricity and clean air, they'll spring for clean air."[50]

People in developing countries, of course, don't choose between clear air and electricity. The problems that lead to the absence of both are obviously more complex than just a decision between one *or* the other. According to Pinker, Americans, Europeans, and Australians have perhaps "sprang for clean air." We are meant to understand that they have "prioritized environmental purity" over economic growth. And in Pinker's mind, poor Africans, Latin Americans, and Asians haven't "turn[ed] their thoughts to the environment." They are surely polluters in Pinker's world. But, for the Harvard professor, that's only a momentary condition. Give them the chance to boost their GDP and they will become as eco-conscious as Scandinavians!

There are many global-data graphs in *Enlightenment Now*. But it seems Pinker forgot to read the one that compares greenhouse gas emissions in rich and poor nations. As it turns out, the richest half of the world's countries emit 86 percent of global CO_2 emissions. The other half (the ones Pinker believes haven't "turn[ed] to the environment") are only responsible for the other 14 percent.[51] Although China, India, and Indonesia are heavy polluters, they produce significantly fewer emissions per capita than the United States, Britain, France, and Germany.[52] And rich nations not only produce more greenhouse emissions, they also consume more water, and generate more plastic and waste (not to mention that they export warfare, fuel conflicts in the Global South, and establish global rules that deprive billions).

Now, back to the 2006 debate: In the other corner was William Easterly, a professor at New York University. Easterly had just published a book with a controversial subtitle: *The White Man's Burden: Why the West's Efforts to Aid The Rest Have Done So Much Ill and So Little Good.* Easterly's cause was joined by Dambisa Moyo, a Zambian economist who wrote a book with an even darker title: *Dead Aid.*[53]

Easterly argued that aid is often based on centralized schemes designed by overconfident decision makers who have little knowledge of what is needed in poor countries or how to obtain it. For Easterly, the aid industry relies too much on humanitarian personnel who deploy preconceived, idealistic ideas. Easterly groups decision makers into two types: planners and searchers. He certainly does not bother with nuance.

Planners, he argues, usually fail to discover the actions that effectively improve people's lives. Searchers, instead, succeed in finding solutions. Why? Because they deploy behaviors that are often found in . . . free markets!

In Easterly's world, searchers respond to incentives and are accountable for their actions. They do not resort to ideologies, and they make rational decisions in pragmatic and systematic ways. For him, these behaviors—and not just aid per se—lead to effective development. Easterly is an unapologetic defender of markets. He contends that "successful businessmen are Searchers, looking for an opportunity to make a profit by satisfying consumers." And then he jumps to show "some of the helpful changes that can happen in aid when accountability is increased, shifting from Planners to Searchers."[54]

I doubt Easterly is followed by the Bonos and Angelinas of this world. But he is not alone in his corner. For most critics of aid, too much money is wasted in international initiatives that rarely produce positive long-term effects and sometimes even create more damaging outcomes. Many critics argue that aid (notably, international assistance) fosters poor countries' dependency on rich ones. Aid, they claim, is largely controlled by political agendas; it is dictated by donor nations, and it often feeds on forms of neocolonialism. Besides, they deplore that aid focuses too much on technology transfer and bypasses legitimate governments and authorities.[55] Criticizing the vulnerability approach to disasters and Western and Northern perceptions of realities in the Global South, Gregory Bankoff argues that between the seventeenth and early twentieth centuries, the prevailing discourse was about "tropicality," and Western intervention was known as "colonialism." "Post-1945," Bankoff writes, "it was mainly about 'development' and Western intervention was known as 'aid.' In the 1990s, it was about 'vulnerability' and Western intervention was known as 'relief.' "[56]

A DEBATE WITH ALTITUDE AND ATTITUDE

This debate, as largely animated by American and European academics and observers, has certainly reached a wide audience. But it has also drawn a simplistic portrait of reality and suffers from a distinct set of problems.[57] The first one is astounding paternalism. Condescending approaches are often found in both defenders and critics of aid. On both sides of the discussion, poor nations are typically portrayed as being dirty and polluting because people in these countries haven't reached a certain level of development.

Experts on either side still claim that environmental consciousness comes with economic growth—an argument that has yet to be proven.[58]

The second problem is that the debate has largely focused on expected impacts rather than real ones.[59] As it turns out, Easterly and several critics of aid have placed too much emphasis on incentives, rational decision making, and other behaviors economists find (or hope to find) in capital markets. Sachs and defenders of aid—here, you may want to include the Gates, Buffetts, Bezos, and Clintons of the world—have placed exaggerated confidence in charities and international cooperation programs.

The third problem is the astonishing lack of autocriticism and nuance on both sides of the debate. Defenders of aid often romanticize the work of charities and international cooperation agencies, failing to recognize their abuses, limitations, and common mistakes. They also idealize donors, overlooking some of their hidden agendas, selfish interests, and partisan ideologies. When confronted about the limits of aid and issues in the work of charities, many defenders of aid respond that such criticism is just an excuse to avoid making sacrifices to help others.

Critics of aid are guilty on this account, too. They demonize foreign aid, global charities, and the work of international agencies, while failing to recognize their contributions to noble causes that produce positive transformation in vulnerable communities. In many cases, they indiscriminately accuse all foreign charities of neocolonialism and imperialism, overlooking the variety of agendas, strategies, and missions that can be found in this multimillion-dollar sector. These critics of aid tend to deplore inequality and oppression, citing reports by Amnesty International, Oxfam, the Red Cross, Greenpeace, and other organizations that—ironically—do their job *with* aid money. They see conspiracies and dubious schemes almost everywhere. When examples of the positive impacts of aid are presented to them, they perceive it as anecdotal evidence and argue that they can't be generalized or scaled up.

Given the polarization of this debate, even economists have found holes in arguments on both sides. Abhijit Banerjee and Esther Duflo, Nobel laureates and cofounders of the Abdul Latif Jameel Poverty Action Lab, are part of this group.[60] They have challenged Easterly's cynicism and the "poverty trap" principle, noting that while many vulnerable residents do, in fact, find it difficult to escape poverty, there are also millions of slum dwellers in Brazil, Mexico, China, India, Chile, and Peru whose descendants have become middle-class citizens in a matter of one or two generations. Banerjee and Duflo instead argue for more rigorous methods to evaluate aid interventions. And they're right: more accountability and rigor are needed when managing aid distribution worldwide. Only by measuring results can we know which interventions work and which don't.

The fourth problem is that both sides of the debate have focused on economic growth as the primary objective and real measure of success—not at all surprising given the panelists' neoliberal leaning. They have both provided a limited explanation of effectiveness, often reduced to a rapid cost-benefit analysis.[61] Most have adopted a simplistic view of value, typically measured in terms of economic growth, profits, and savings. Even Banerjee and Duflo are predominantly concerned with economic indicators of success. They are interested in aid's capacity to improve income, savings, diversity in resource generation, and of course, profit and growth for entrepreneurs. But both sides of this discussion have repeatedly failed to acknowledge that, in a world of unnatural disasters, economic growth is an insufficient measure of the change we need. Other thinkers, such as Amartya Sen, Martha Nussbaum, and Michael Sandel have recently reminded economists and political scientists that there are many other forms of value. These nonmonetary benefits include freedom of choice, honor, sense of belonging, and the comfort of traditions and collective narratives. In chapter 8, we will see how all of them contribute to our fight against climate change amid an increase in frequency and intensity of disasters.

The fifth problem is that both sides of the debate have concentrated on the impact of aid alone, failing to see the bigger picture that informs relationships between rich and poor countries.[62] Several studies have proven that aid is only a small fraction of the overall framework that rules relationships between the Global North and South. One example of this is the influence of aid on development. At first glance, figures highlighting the amount of aid given to poor countries appear impressive. But a different perspective emerges when one looks at the bigger picture. Recent studies have shown that aid does not have a major impact on many poor countries' budgets. In Africa, for instance, foreign aid funded under 6 percent of total government budget expenditures in 2003.[63] In Kenya, only 5 percent of the national budget comes from foreign aid.[64]

Too much weight has been put on aid in the relationship of power that exists between rich and poor nations. Hickel has found that financial aid flowing from rich to poor countries (estimated at $128 billion in 2017) is relatively small compared to the flow of capital from poor to rich countries; the "aid budget turns out to be a mere trickle," he says.[65] This flow of resources takes different forms. One of them is international debt, paid by developing countries to rich ones. Another regards the profits that foreigners make on their investments in poor nations, which they repatriate. Others include the payments to patent owners in rich countries. This flow also includes what Hickel calls "capital flight," which involves a variety of mechanisms of corruption used in international trade.

Hickel notes that this reality is quite counterintuitive. We tend to believe that rich countries help the poor, not the other way round. But he carefully shows how policies of trade, international agreements, loans, copyright regulation, and political manipulation have led to poor countries being oppressed by the rich ones. Many poor nations do not have the capacity to compete or oppose the rules and conditions set by Europe and the United States. That's how they end up losing massive amounts of money to rich nations, and—ironically—only receive a small fraction of it back in the form of charity. He calls this phenomenon "aid in reverse."

The final problem—and probably the cause of all the others—is that this debate has historically been conducted at a view of thirty thousand feet off the ground. It feeds on generalizations, assumptions, and abstractions.

Geopolitics today are not limited to a struggle between the North and the South, democracy and communism, the West and the Rest. I have seen relationships beyond these binary distinctions shape disaster-related aid in so many places. A great deal of international aid occurs today within countries of the Global South. In Haiti, I visited postdisaster tents that had been shipped from China, and residential projects built with money raised in Colombia (see figure 2.2). In Cuba, I visited houses built with petrodollars from Venezuela. In South Africa, I worked with engineers provided as "aid" by the Cuban government; and millions of Cuban doctors work today across Africa and South America.

FIGURE 2.2 (*Top*) China is active in providing emergency aid. This Chinese shelter was sent to postearthquake Port-au-Prince. (*Bottom*) Aid is now commonly sent between countries in the Global South. This housing project in Haiti was built with money collected in Colombia.

Source: Gonzalo Lizarralde, 2014.

Arab countries provide aid based on religious and political agendas that do not correspond to the conventional binary relationships that economists explored in the 2000s. Venezuela provides significant aid to socialists and left-wing governments in Central America. China is channeling the equivalent of millions of dollars in aid to African and Asian countries.[66] Charities from South Africa, El Salvador, Chile, Brazil, Ecuador, and many other low-income nations have international activities that are not ruled by traditional rich-poor geopolitics. Turkey, Japan, and South Korea sent about 1,400 prefab shelters to victims of the earthquake that hit Bam, Iran in 2003.[67] A similar trend now occurs in international money lending. A 2019 report revealed that China is now the world's largest creditor.[68] China's overseas lending to mainly poor countries went from almost nothing in 2000, to more than $700 billion today—more than twice the World Bank's and IMF's combined.[69]

Another common generalization is to place all charities in the same group. Just as some NGOs denounce abuses in ideology and faith-oriented policies, others support these same principles. Some relatively small NGOs emerge from within communities, adopt simple and horizontal structures, and represent their own members. Others are created within religious congregations and have a spiritual mission. Still, there are NGOs with the same sophisticated hierarchies that characterize corporations. And there are also a variety of multinational charities, that sometimes work as franchises or federations, with satellite organizations in different countries and regions. These, and many other types of charities and agencies, have diverse work practices, strategies, and underlying missions.

In this book, I try to focus on the realities I have seen not from thirty thousand feet up but among disaster-affected communities and real people on the ground. I believe that when the analysis is taken to the ground, a less polarized perspective, with more nuances and shades, emerges. A closer view of activities on the ground also reveals how disaster capitalism influences reconstruction and risk reduction.

At the ground level, I have found a different set of concerns. Emergency aid officers, for instance, routinely report that nonaffected people pose as victims to claim assistance and resources. Too many people donate the wrong kinds of stuff after disasters (such as rotten food or video games for devices that are not readily available). Local politicians and intermediaries often steal money and resources intended to help disaster victims. Humanitarian solutions are frequently fragmented, and once donations are received, miscommunication with the public is common. In general, charities lack rigorous management and efficient logistics.[70]

But perhaps the most harmful problem I have witnessed is the lack of understanding of beneficiaries' real needs and expectations. To be clear, this problem is not unique to charities engaged in disaster *reaction*. I have also seen it in those focused on *preventing* disasters. In a world of unnatural disasters, there is surely room to contest the effectiveness of aid and simultaneously denounce the influence of savage and disaster capitalism. But we must also challenge the values that motivate donors, charities, and international agencies. To do this, let's first explore the role that aid plays in disaster reduction and the quest for sustainable development today.

FROM EMERGENCY BLANKETS TO SUSTAINABLE MASTER PLANS

In my twenty years of work in the disaster field, I have witnessed a major shift in the scope of aid. When it comes to disasters, charities had traditionally concentrated on emergency aid (for instance, the Red Cross originated in Geneva in 1859 when a young Swiss man set up an organization to heal soldiers' wounds and feed and comfort them during war[71]). Several relief organizations worldwide, including the Emergency Architects in France, also started as first responders during wars or major disasters. They normally provided technical assistance, medical support, bed nets, food, clothes, blankets, and hot meals. But decision makers in funding agencies and charities have noticed four significant realities within their sector.

The first reality is that temporary (and thus substandard) solutions, produced in times of emergency, frequently become permanent. Refugee camps, for instance, often become neighborhoods, which then enlarge the urban footprint permanently.[72] Emergency shelters are rarely ever dismantled. After earthquakes took place in Iran and El Salvador, the temporary shelters were later used by residents as storage rooms or additional bedrooms tacked on to their permanent houses.[73] In some cases, these shelters are even traded in the local real estate market. Temporary pieces of infrastructure, such as wells or prefab facilities, are also left in place far longer than initially expected.[74]

The second reality is that so-called solutions often have a limited, short-term effect, failing to impact people's lives in more structural ways. In many cases, Band-Aid solutions actually cause long-term *negative* effects. Take the case of free medical aid provided by foreign charities. In 2016, one of the members of my team in Haiti got injured and we had to take her to a clinic in Pétionville. There, I met a charismatic and very professional Haitian physician. He told me that business was difficult at the time. "We can't compete with foreigners who offer free consultations and exams," he explained. The physician believed that free foreign medical aid, provided in the past few years in Haiti, had caused many bankruptcies among the private clinics run by local doctors.

The third reality is that charities stay in vulnerable areas only for short periods of time. Most NGOs shift their attention quickly, chasing opportunities for fundraising wherever disasters, famines, and refugee influxes are making headlines. Almost all charities I saw in Port-au-Prince in 2010 had left by 2015. Some took up work in Nepal, after an earthquake partially destroyed the historic city of Kathmandu. Most charities, therefore, fail to develop the long-term relationships with local stakeholders and communities that are required to build trust, find appropriate answers, and develop effective partnerships.

Finally, decision makers discerned another reality—disasters are ideal opportunities to raise money. Given the amount of money collected in times of disruption, an increasing number of organizations now wonder: Why stop at providing blankets, bed nets, and food? Why build a temporary clinic when we can build a permanent one? Why set up tents when we can build houses? Charities and international cooperation agencies thus have grown more ambitious.

Today, more and more nonprofit organizations are embracing the objective of planning sustainable cities and building green residential neighborhoods, infrastructure, and services. In many cases, NGOs are no longer providers of temporary aid but designers of permanent, often irreversible, solutions. Given the problems caused by short-term solutions, this shift might look like a positive one. But as we shall see, it comes with its own troubles.

CHANGE AND HUMANITARIAN SUSTAINABLE DEVELOPMENT

I met Nancy in Cape Town in 2006. A kind and generous person, she was about twenty-five years old when we met. She had just graduated from architecture school in Ireland and was working with an Irish NGO involved in low-cost housing in South Africa. She was eager to apply the knowledge and skills she had gained at school. She had visited the townships (slums) in the Cape Flats, where she witnessed poverty and the impact of apartheid on Black and coloured communities. She wanted to help to redress these historic injustices, and her job gave her that opportunity.

Nancy worked for the Miller Residential Initiative,[75] a charity created by Irish multimillionaire Nestor Miller. Miller had amassed wealth in Ireland through a successful career as a property developer. He wanted to give back to those in need, so he set up a nonprofit organization to build sustainable solutions for the poor. The charity partnered with a South African NGO to build permanent houses for marginalized residents in the townships. Most of the beneficiaries were survivors of urban fires, which had become frequent in the informal settlements around Cape Town.

If Nancy had stayed in a wealthy country like Ireland and worked as a junior architect in a firm, she probably would have joined a design department. She would have worked under the guidance of a senior architect with extensive experience in housing design and construction. The architecture team would have included experts in construction codes, building specifications, 3D modeling, technical details, municipal regulations, green construction, and other important tasks required to produce high-quality residential buildings. Other professionals in engineering, urban planning, product design, sociology, and anthropology may have reinforced the project team. Nancy's role would have probably been modest in the overall project structure. As a young professional, she would have likely supported more senior architects and urban planners as they worked out specific aspects (construction details, for instance) of housing plans. But not in South Africa.

In the developing-country/not-for-profit version of Miller's construction company, Nancy was appointed *the main architect* of a 192-unit residential development. The project was aimed at helping some of the most vulnerable South Africans living in Bonteheuwel.

The original plan was to build the houses with Irish and local volunteers. But like most low-cost housing initiatives, the Bonteheuwel project was socially complex. It centered on a community that had fallen victim to one of the most brutally racist systems of rule after the Second World War. And like many other residential development projects, it was technically difficult. Designers had to make changes in zoning and soil treatments, plan new roads and drainage systems, and coordinate an intricate financial plan involving transfers of land titles. The project also necessitated a water collection system and the construction of buffer zones to shield homes from pollution, among other detailed needs. All of this was too sophisticated for unskilled volunteers, so construction companies were hired.

In light of this challenge, Nancy did her best in all regards. To make up for her lack of knowledge and experience, she put in extra work. But leading such an ambitious and complex project—from all social, cultural, political, and technical points of view—was overwhelming, and the result was disappointing.

The new neighborhood in Bonteheuwel was needlessly low-density: it had a reduced number of one-story homes in large plots and too much space devoted to roads. Due to its configuration and construction, houses were difficult to upgrade and upsize over time. Open space was wasted and the poor urban design further increased the fragmentation of the neighborhood. Housing types did not correspond to common local practices, most notably the way that women in the South African townships conduct domestic and social activities. When women cook or undertake chores, they are often facing the road—a strategy that contributes to supervise public space and increase security. In the new houses, kitchens and sinks were placed at the back of the plot, so women lost the connection with the street that characterizes their vernacular solutions. Worst of all, the design prevented residents from performing home-based income-generation activities. As such, many of the "beneficiaries" lost the home-based businesses they had run in their previous locations (often shops and workshops). By losing these crucial sources of income, many of them became more vulnerable after moving into the new houses.

A few years later, the Miller Residential Initiative was dismantled. Miller was accused of fraud in South Africa due to his foundation's involvement in other dubious activities. He was later convicted of public drunkenness in Dublin. His company went bankrupt (it is estimated that Miller lost €150 million in the Irish property crash). But Miller didn't give up. In 2004, he set up a new charity to improve education for South African children.

In my work, I regularly meet well-intentioned professionals, like Nancy, who want to apply their skills in situations of crisis. Most of them see suffering and believe that they can help alleviate it. Many find causes that match their motivations, and today, hundreds of organizations represent their interest in sustainable action. These organizations are often named after a combination of professional titles and social causes: Doctors without Borders, Engineers for Humanity, Town Planners without Borders, Emergency Architects, architects that design "like [they] give a damn" (the title of a popular book on the subject[76]), and many more. We should celebrate these professionals' commitment to helping others. But producing positive change requires more than good intentions and fluency in sustainable development rhetoric. Many employees in the aid industry think that they are pedaling in the right direction, when the vehicle is veering wildly off-track.

UNSUSTAINABLE OVERHEADS, UNSUSTAINABLE QUALITY

The Irish charity that employed Nancy did not apply the same rigorous methods that the construction company that birthed it had likely applied in Ireland and Europe. Instead of creating a multidisciplinary team capable of dealing with the complexity of the task, it engaged a recent graduate with good intentions but limited expertise. Unfortunately, this is not rare. The Bonteheuwel case illustrates three common problems that exist today in the aid industry.

The first problem is, perhaps unsurprisingly, lack of expertise. Plenty of professionals have the skills needed to produce positive long-term change, but charities rarely engage them, nor do they expend many resources creating the multidisciplinary, experienced teams required to deal with intricate projects for vulnerable populations. The main reason for this is that a sophisticated team with experience and expertise is expensive. And, in the realm of aid, administrative costs are often kept to a minimum.

Charities are pressed to reduce overhead and administrative costs. The reason lies in the very nature of the aid industry: when donors give money to charities, they want their money to be used "on the ground." They want their donations to reach beneficiaries directly. When, say, Rebecca, a woman in Dublin, donates $100 to build an eco-friendly house for Patrick, a child in Cape Town, she wants her money used to build Patrick's home. She probably does not want to pay the handsome salary of an urban planner in Johannesburg or Pretoria. Rebecca and most donors are rarely interested in giving money to pay for staff, administrative costs, and overhead. Charities are, therefore, under constant pressure to reduce the costs of design, planning, and management. A way of doing this is to hire a few young, enthusiastic professionals like Nancy. This is less costly than hiring large teams of senior experts. This also explains why charities are also often staffed with volunteers. Unfortunately, while these volunteers might have the best intentions, they lack the advanced skills required to produce long-term, structural solutions.

The second problem lies in the mandates usually adopted by charities. Again, this is an unfortunate consequence of the very nature of aid. Charities obviously have to motivate donors to give money, and as such, try to propose projects that will paint vivid images in donors' imaginations. In reality, building houses was probably not the best way to help the Bonteheuwel community: giving cash to entrepreneurs in the informal sector might have had a more positive impact on beneficiaries' lives, just as building a sewage system would have been better for the environment. But it is easier to collect donations for building "green" houses that can be beautifully depicted in sustainability reports, brochures, and websites, than it is for cash transfers or sewage lines. In the case of cash grants, donors might complain that they won't know how poor people will spend the money (one officer once asked me: What if beneficiaries spend the cash on beer or prostitutes?). As for sewage, I suspect that pictures of pipelines and reservoirs of fecal matter are not sexy enough to attract donors.

Nancy's motivation and energy could have been directed toward monitoring the use of cash subsidies among women entrepreneurs over a period of five to ten years. This could have provided valuable insight for improving people's lives. Or, she could have supervised the installation of badly needed sewage pipes. But instead, she was hired to produce houses that, in some cases, had a limited positive effect, and in others, destroyed the very livelihoods of their "beneficiaries."

The third problem is linked to the value of design. Another reason why design-oriented charities often fail lies in their objective: they seek to apply professionals' pre-acquired skills in situations of fragility. These professionals rarely change their attitudes and techniques when confronted with the complex social and environmental challenges created by risk and vulnerability. As aid officers, they try to design well-intended solutions, such as green neighborhoods, clinics, and schools in slums. Yet they typically deploy their *existing* skills in the process—the same abilities they would use to design a LEED-certified corporate building. Some of them eventually realize that their solutions do not produce the expected results. A few of them do engage in cultivating the type of skills that are *actually* required to face these challenges (which I will explore in the following chapters). More often than not, however, this shift occurs only after several mistakes are made, and most of these errors have permanent consequences.

Design is a form of power that is magnified in contexts of vulnerability. Misuse (even with good intentions) can have disastrous results. Contrary to what most of my designer colleagues think, I believe there are two significant differences between design in general and design in contexts of risk and vulnerability. One difference is *consent*. Designers tend to make products and spaces that are used and frequented by individuals voluntarily. If you do not like a chair designed by Frank Gehry, you are not obliged to buy it. You can also choose whether or not to stay in a hotel designed by Philippe Starck or visit the experimental interior of a certain museum.[77] But consent is not so easily exercised in the public spaces of vulnerable communities. Citizens can't simply refuse to see a public facility or use the latrines in their neighborhood. Neither can they decide whether to use a public space to conduct civil duties or cultural rituals. Buildings, public spaces, and infrastructure must be seen and used regardless of whether their users consent.

Another difference is the role that design plays for vulnerable people. For most wealthy and privileged individuals, designed products of consumption shape their *private* lives in leisure activities, such as exercise, as well as income generation. But planning and design schemes in the face of risk play a fundamental role in citizens' *public* lives. The design of cities and human settlements determines power relationships between citizens and their public institutions, which helps to build or erode people's sense of belonging. Infrastructure, housing, and services largely inform their choices, enhance or restrict their rights and freedoms, and affect their chances of leading fulfilled lives. Contrary to a belief held by most of my colleagues, design and planning in the face of risk does *not* have intrinsic value.[78] In fact, disaster reduction and response efforts feature as many examples stressing the benefits of design as instances of its abuse by the most powerful.

CHANGE IN A WORLD OF UNNATURAL DISASTERS

I have found that the aid industry is filled with well-intentioned, committed, and hard-working people like Nancy. But it has structural problems that need to be addressed. Some of them require a high-level perspective, whereas others require further attention at the level of operations, logistics, and implementation. But these changes won't be possible without a serious debate about the value and ethics of giving and protecting. To do this, we can't rely on prefabricated, off-the-shelf ethical principles based on the catchphrase of sustainable development. We can't assume that there is an intrinsic value in humanitarian actions or sustainability initiatives. Instead, we must constantly and rigorously challenge the value of our actions. We must question the moral worth of specific interventions, and carefully select the causes that we support with our time, money, and attention.

The problems currently faced by the aid industry remind us how difficult it is to identify what disaster victims and vulnerable people want and need, and how to respond to their expectations. In a world of increasing unnatural disasters, different forms of aid will be required. We need to distinguish among different types of solutions—particularly among those that:

- Are incremental, and leave room for evaluation, testing, and improvement vs. the radical plans and schemes that leave little space for adaptation and perfection.
- Emerge locally, solving specific problems and needs within their own time and context vs. those built on oversimplified approaches and broad assumptions.
- Are identified through honest dialogue between parties and are capable of creating consensus or compromises vs. those that are imposed from outside by people pursuing partisan agendas.
- Are built on shared values and collective narratives vs. those that simply seek to recreate the competition and inequalities that characterize markets.
- Recognize the significance of local modes of living, values, and struggles vs. those that reproduce forms of colonialism and imperialism.
- Support vulnerable people's own causes vs. those that are built on unverified assumptions about their needs and expectations.
- Listen to the voices of vulnerable people vs. those that impose a foreign narrative on them.
- Condemn social injustices vs. those that perpetuate them.

It will be difficult. There is no magic trick for accomplishing this, except for a constant and arduous examination of the value of our own actions. Future disasters will test our capacity for compassion, empathy, and self-criticism. They will challenge our assumptions about the type of change that is needed. They will also reveal the differences between us, our cherished values, and our needs and expectations.

So far, there have been several disconnections between the type of change that is imposed and the one that is actually needed. Global warming is producing irreversible transformations. We must change in order to avoid even more problems. In the next chapter, I will argue that, by invoking the catchphrase "sustainable development," people in power have found a way to pretend that they care about nature, as though sharing in the interests of common citizens.

3 | Sustainability

"They Often Come Here With Their Talk About Green Solutions"

The fair and reasonable desire that all men and women, rich and poor, be able to take part equally in the Lottery inspired indignant demonstrations—the memory of which, time has failed to dim. Some stubborn souls could not (or pretended they could not) understand that this was a novus ordo secorum, *a necessary stage of history.*

—Jorge Luis Borges, "The Lottery in Babylon"

FLOATING ON AN OCEAN OF IDEALISM

"Welcome to Harvest City," begins the promotional video. "A two-mile wide floating structure that sustainably expands the country's inhabitable and productive area, into the Caribbean Sea."

Harvest City is presented as a solution to the destruction caused by the 2010 earthquake in Haiti. The idea is to build a new city on the water of the Port-au-Prince Bay, solving the housing and infrastructure deficits of the capital city. This concept not only includes hurricane-resistant floating units for thirty thousand residents but also floating farms, a floating harbor, markets, schools, and other facilities. "Through a network of roof-top solar collectors and perimeter wind turbines," state the designers, "the city will produce a large amount of its own energy." A system of boats and canals will guarantee transportation to and within the floating structures.

Launched in December 2010, Harvest City was never built. The concept captured minimal attention on the web (a video is available[1]) and likely none among investors. Harvest City is one of the multiple utopian futures that emerged after the 2010 disaster. Its futuristic and 3D animations might not deserve much attention, and yet Harvest City was not designed by amateur futurologists. It was the vision of Schopfer Associates LCC, a serious, experienced, and award-winning architecture and planning office based in Massachusetts. The narrative surrounding Harvest City is not idealistic, but it is full of the rhetoric commonly used by mainstream sustainable developers today. In one sense, Harvest City demonstrates the naïvete and sense of superiority that characterize many of the designers who believe to have found the solution to a postdisaster context. Yet the floating houses, neighborhoods, and cities are increasingly considered effective solutions to climate change effects.

The disconnect between consultants and other ideologists in the Global North and the realities of low-income, vulnerable populations in the Global South are not rare. Nor are they fully surprising after decades of colonialism, neoliberalism, and imperialism. It is perhaps more surprising though that a significant disjunction also exists between pragmatists in the Global South and the realities of their own low-income, vulnerable populations.

ALL I WANT TO SAY IS THAT THEY DON'T REALLY CARE ABOUT US

Say you want to move to a favela (the Brazilian term for slum). How much would you be willing to pay for a house in a favela in Rio de Janeiro?

Anything less than US$60,000 probably won't cut it. Ms. Do Santos was asking $60,000 for her home in Favela Santa Marta when I visited her in 2009. She wanted to take advantage of the gentrification process occurring in Santa Marta and in other Rio de Janeiro favelas. In 2009, Ms. Do Santos worked as a cleaning lady and sold clothes to make an extra income. In many ways, her home was probably not what many people would expect to find in a slum. It was a small but lovely two-bedroom house with a magnificent view of the beaches and mountains that make Rio a truly special place. It was a well-furnished and functional home. Inside were two large TVs, a new computer, and a giant sound system. Ms. Do Santos had just painted the façade. Her plan was to sell her home and move with her two children to Rocinha, a notorious slum located farther south.

A devoted mother, Ms. Do Santos is not the type of outspoken social activist you would generally find in the favelas. She is instead reserved and somewhat shy. But if you are a fan of Michael Jackson, you have probably had a (no doubt, cartoonish) glimpse of the social activism scene in Favela Santa Marta years ago—and you might have even seen images of the neighborhood where Ms. Do Santos lived. The favela was one of the locations where Spike Lee filmed Jackson's performance of "They Don't Care About Us." When the King of Pop danced in the favela, Santa Marta was one of the most dangerous places in town. Yet, by 2009, it was rapidly transforming. The public authorities were working on a comprehensive upgrading plan that included the construction of two police stations, a garbage collection system, a network of sidewalks, a nursery, public squares, a public cable car railway, and—since we are talking Brazil here—a football pitch.

Santa Marta has always benefitted from a strategic location close to jobs and services. After the upgrading program, hundreds of local residents viewed it as an attractive place to live, in spite of the increasing rents. It integrated the favela with the neighboring and more affluent sector of Botafogo. With urban integration and better mobility, home-based businesses in the favela thrived. Family incomes and a sense of tenure security increased. Residents invested in their homes, engaged in additional construction activities, and improved their structures and façades. Many, like Ms. Do Santos, sold their homes to more prosperous residents and moved to other favelas offering cheaper homes. A form of gentrification settled in.

But the slum-upgrading program had an additional feature that would become even more controversial than the rapid change of social status: walls.

For years, *favelados* (slum dwellers in Brazil) had been frequent victims of crime, pollution, floods, and other hazards. Landslides in the steep slopes occupied by informal housing were common in Santa Marta and in other slums, particularly after heavy rains. It is estimated that between 1990 and 2010, rain-related disasters affected nearly five million Brazilians.[2] In 2011, landslides caused by persistent rains killed hundreds of people in areas close to Rio.[3]

Deforestation has also been rampant. Environmentalists estimate that Brazil "lost an average of 2.9 million hectares of forest per year between 1990–2000 and 2.65 million hectares per year between 2000–2010."[4] In 2019, the world was outraged by news of uncontrolled fires in the Amazon forest—hundreds of them.[5] Most deforestation in Brazil is caused by subsistence and commercial agriculture, infrastructure projects, logging, mining, and cattle ranching (Brazil is the world's largest producer of beef). Fires are a common source of destruction not only in the Brazilian Amazon but also around cities. Between 1991 and 2000, there was an average of seventy-five forest fires per year in Rio de Janeiro. Many of them partially burned the National Park of Tujuca, the largest urban forest in the region.[6] When the sprawl of Santa Marta and other central favelas threatened the protected green areas and scenic mountains behind the beaches of Rio de Janeiro, the authorities were truly worried. With the Olympic Games scheduled to begin in 2016, protecting the forest near the beaches from informal construction became a top priority.

The authorities had noticed that slum communities were growing eight times faster than other areas of the city.[7] They claimed that slum sprawl had become one of the main causes of deforestation in the state of Rio de Janeiro—an argument that was exaggerated, or simply inaccurate, given the historic effects of fires and agriculture. It is true that landslides were systematically destroying new makeshift homes in sloped areas and killing hundreds of favelados. Additionally, hundreds of families with US$60,000 in their pockets were willing to move to Santa Marta to experience the view and proximity to some of the jobs and opportunities that Ms. Do Santos had enjoyed for years.

In response, the authorities launched a US$17 million plan to build walls that would contain Favela Santa Marta and a dozen more of the hundreds of slums that exist in Rio. They believed that constructing four metre–high concrete walls around these slums would discourage construction on their peripheries. Santa Marta was the first test site for the concrete wall program.

The plan stirred a heated debate. Critics argued that the money devoted to walls could be better used to foster environmental awareness or to coordinate a community action plan for protecting the forests. Many contended that the real concern of the authorities centered on the image that Rio had to promote during the Olympic Games. They saw the walls as a way to hide the shantytowns from tourists. Analysts noted that if the walls managed to prevent new construction in the favelas, informal settlers would simply build their houses in other green areas. Low-income families would instead settle on the outskirts of Rocinha and in other protected forests.

When I visited the walls in Santa Marta (see figure 3.1), the residents and social leaders were in a state of outrage. They claimed that the walls didn't succeed in preventing urban sprawl. But even if they did, the barriers restricted the rights and freedoms of local residents.

As one resident explained to a CNN journalist, the authorities "treat the people here like their children, who need to be corralled in."[8] Many locals felt imprisoned and deplored the fact that they had been deprived of the views that richer people enjoyed without restriction. Some even created parallels between the new barriers and the Berlin wall. A few months after completion, residents had already perforated the wall in several places, an effort to prove that the barrier could be easily dismantled by community action.

The "eco-wall" around Santa Marta became a symbol of the segregation and exclusionary politics that have marked Brazil's social structures for decades. The plan for building walls in other favelas was eventually abandoned. Meanwhile, the unchecked deforestation in the country continued. Brazil lost 6,624 square kilometres of forest in 2017 and 10,000 in 2019.

The walls that were built in the mountains of Rio were an extreme solution to an environmental challenge. Yet this case illustrates a common trend today: neither experts nor ordinary people like us agree on how to effectively protect nature or how much to pay for that protection. It is also an example of the type of guilt-erasing measures we often adopt. We can refer to them as "the plastic-straw fallacy"—that is, the tendency to concentrate

FIGURE 3.1 (*Top*) The view of Rio de Janeiro with Favela Santa Marta in the foreground. (*Bottom*) The eco-wall built around Favela Santa Marta.

(Republished with permission of Taylor & Francis from Gonzalo Lizarralde, *The Invisible Houses: Rethinking and Designing Low-Cost Housing in Developing Countries* [New York: Routledge, 2014], permission conveyed through Copyright Clearance Center.)

our environmental protection efforts on measures that require minimum sacrifice but offer abundant visibility, such as banning straws. This happens when we focus on symbolic measures rather than on the structural changes that demand real sacrifice and compromise.

The problem is that measures taken under the plastic-straw fallacy, like building walls to prevent deforestation in a particular section of Rio, usually have little impact. While they tend to become successful public relations campaigns, they rarely bring us any closer to reducing our impacts on the environment.

They also distract public attention from the real solutions we need and the urgent debates we must hold. Quite often, they also create dramatic secondary effects, such as the feeling of imprisonment favelados have experienced.

At one point or another, we have all fallen into the trap of the plastic-straw fallacy. This stems from an umbrella of ideas put forth by industry and our governments—a term that is United Nations-sanctioned and now recognized worldwide: "sustainable development."

OUR (REMARKABLY POPULAR) COMMON FUTURE

> *Economic growth and development obviously involve changes in the physical ecosystem. Every ecosystem everywhere cannot be preserved intact. A forest may be depleted in one part of a watershed and extended elsewhere, which is not a bad thing if the exploitation has been planned and the effects on soil erosion rates, water regimes, and genetic losses have been taken into account.*

If you have been following the catastrophic effects of climate change and are as alarmed by them as I am, you might believe that the previous paragraph was written by the CEO of an international oil company. Shell? Exxon Mobil? Or, perhaps the CEO of an agrochemical company—Monsanto? You probably think that this CEO is so ignorant that he (I have reason to believe that you're imagining a male leader) views the trees in a forest much like inanimate objects that you can eliminate "here" to put "there." You might deplore him for failing to realize that a tree is *a living creature* and not a piece of furniture that humans can relocate on a whim. If you know a bit about watersheds, you understand that they keep us—you and me, as well as millions of other living species—alive through a constant supply of water and essential nutrients. As such, you might be outraged by this CEO for thinking it is okay to deplete a part of a watershed "here," so long as some trees are planted "there."

What would you say if I told you that this text was *not* written by the CEO of a polluting multinational business but by the most influential committee ever mandated to set guidelines for protecting nature now and for generations to come? The committee, led by a woman, for the record, was notably backed by the United Nations.

As it turns out, the quotation at the beginning of this section is extracted from what came to be known as the Brundtland Report.[9] This is a document that, in 1987, popularized the concept of sustainable development worldwide. It was written by a committee led by Gro Harlem Brundtland, a politician who served three terms as prime minister of Norway. The document, also called *Our Common Future*, set up guidelines for achieving environmental protection while accelerating economic growth on a planetary scale. The quoted paragraph illustrates just how low the bar was set in 1987.

After Brundtland, sustainability emerged as a collective agreement to "save" the planet. *Our Common Future* was structured in three sections: "Common Concerns," "Common Challenges," and "Common Endeavours," assuming that those things existed. The tone suggests an inevitable compromise between environmental protection and economic growth.

To be fair, the report does promote some sound ideas, such as engagements to solve low-cost housing (quantitative and qualitative) deficits, plans to improve infrastructure and public services in marginalized areas, and the involvement of the informal construction sector in poor countries. It also puts forward an argument that I emphasize in this book: nations must "confront a growing number, frequency, and scale of crises."[10]

At first glance, it is difficult not to see the Brundtland Report as a manifesto of good intentions: the overall principle is sound. Social, economic, and environmental considerations—the three pillars of sustainable development—have to be considered simultaneously to achieve the type of lasting development for future generations. A less glowing view, however, often emerges after more in-depth reading. The underlying message of the report is that businesses should be run in the usual manner to foster industrial development and continued economic growth. In the words of Brundtland and her coauthors, nature is portrayed as a resource for development now and economic growth for years to come.

In spite of this (or—as I shall argue later—*because* of it), sustainability was naturally adopted by industry. A global survey conducted by McKinsey & Company, an American consulting firm, found that 57 percent of business executives in 2011 had already seen their companies integrate sustainability into strategic planning.[11] Similarly, a study conducted in 2018 found that as many as 60 percent of construction-project clients and professionals expected their projects to be green in the near future.

In the same way that your morning coffee can be certified as "fair-trade organic," labels exist today to certify buildings, infrastructure, neighborhoods, and even cities. You may have heard about "leadership in energy and environmental design" (LEED) and the "building research establishment environmental assessment method" (BREEAM). These are the most popular green certifications for buildings in the United States and the United Kingdom, respectively. In the early 2000s, there were only about forty LEED-certified buildings in the United States. Eighteen years later, there were more than sixty-seven thousand LEED-registered buildings[12] and more than ninety-five thousand projects using LEED worldwide.[13] Today, there are also now more than 2.2 million BREEAM-registered buildings in more than eighty countries. And these represent just the tip of the iceberg. It is estimated that there are over six hundred different green certifications for buildings and neighborhoods alone.[14] Green construction is now a multimillion-dollar business worldwide, an industry of consulting services, and an economic sector with rapid growth.[15]

Thirty-two years after the Brundtland Report, sustainability has mutated into a mammoth industry of consultants, university professors, academic grants, inventors, books, courses, graduate programs, industry fairs, certifications, and of course, international conferences (some of which I have attended, burning lots of fuel in air travel, thus contributing to global warming).

During the past three decades, the idea of sustainability seduced consumers worldwide. A global survey conducted by Nielsen, a market research company, found that 66 percent of consumers in 2015 were willing to pay more for sustainable brands—up from 50 percent in 2013.[16] Another study found that millennials in the United States consider it crucial for their purchases to be "inherently sustainable."[17] Similar studies have shown that 77 percent of Americans think that sustainability is a key factor when deciding on what food to buy.[18] It is estimated that as many as 92 percent of Americans have a more positive image of companies that support social and environmental issues.[19]

Sustainability has captured the imaginations of scholars and researchers as no other concept has. Its rise has steadily continued since 1987. Initially, in 1979, only a few hundred academic papers and books had mentioned the term "sustainability." In academia, it was less popular than terms such as "disasters" and "climate change." Within a few years, though, the popularity of sustainability surpassed these other environmentally-related subjects. Sustainability even became more popular than scientific concepts such as black holes and artificial intelligence. This is surprising given that these last two terms entered pop culture through widely admired films like *Star Wars*, *Star Trek*, and other robots-flying-in-the-stars shows (see figure 3.2). Sustainability technology became a profitable business. A study by the European Patent Office found that green-construction-related patent filings tripled between 2000 and 2011.[20]

Sustainability also became widely popular in policy worldwide. In Europe, it dominated both policy and political discourse. In fact, it became so common that, in 2008, Vladimír Železný, a controversial Czech politician and member of the European Parliament, filed a formal declaration on the "overuse of the adjective 'sustainable.' " His motion noted that "the word 'sustainable' is repeated in almost all policies and strategies, producing empty, meaningless phrases that may give rise to dispute." Exasperated by greenwashing, he asked the Parliament to adopt a "temporary moratorium on the word 'sustainable.' "[21]

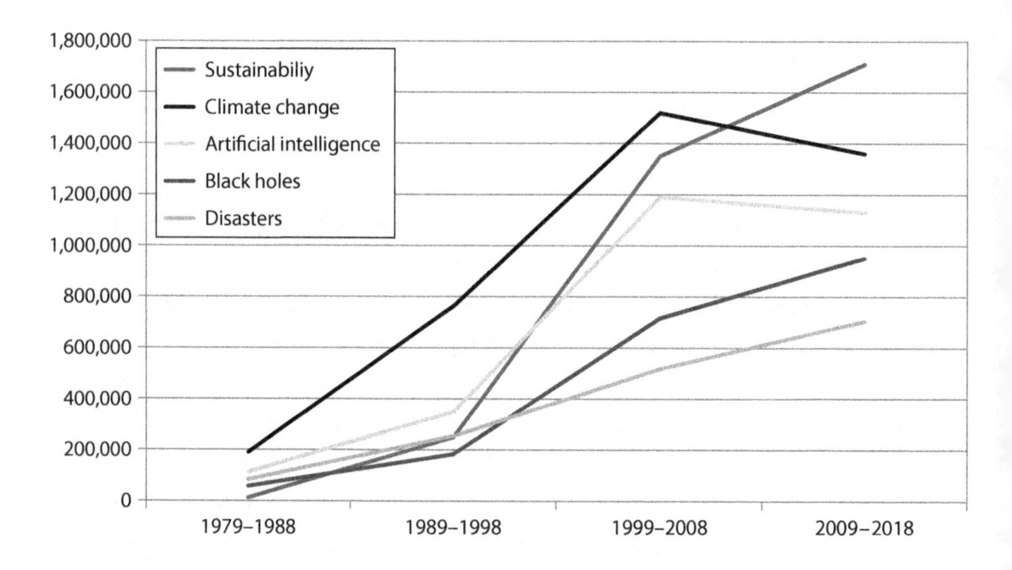

FIGURE 3.2 The popularity of sustainability when compared to other concepts used in academic journals and books. On the vertical axis are the numbers of academic texts citing these terms. (Data obtained from Google Scholar.)

OUR COMMON DECEIVING FUTURE

Since the Brundtland Report was published, many indicators of well-being have improved. Disasters cause fewer casualties than in previous decades. Life expectancy has also increased in almost all areas of the world. Crime and violence have reduced significantly in rich countries, and advances in health and sanitation have spread to many of the poor ones.

But as sustainable development has grown in popularity, many other aspects have worsened. There has been a severe increase in the intensity and frequency of disasters. Between the publication of the Brundtland Report and Michael Jackson's visit to Favela Santa Marta in 1996, humans accelerated the burning of fossil fuels, putting three billion tons of CO_2 in the atmosphere. We then added another thirteen billion tons between the "They Don't Care About Us" video clip and 2018. All of this was enough to increase the global temperature by about 0.4 degrees Celsius,[22] pushing us closer to the level required to trigger climate-induced destruction on a planetary scale.

NASA estimates that from 1987—when the Brundtland Report was published—to 2015, the planet lost about 2.5 million square kilometres of Arctic sea ice.[23] From 1993 to 2018, the sea level rose 80 millimetres—making disasters all too common in coastal areas around the world.[24] Biodiversity also declined: environmentalists estimate that between 1970 and 2014, there was an overall decline of 60 percent in wild species' populations. During that period, the world lost about half of its shallow water corals.[25] By 2007, 29 percent of seafood species were fished to the extent that they could not reproduce themselves.[26] We also put so much plastic in the ocean that we now regularly eat plastic particles in our diet. We thus had to learn about an unprecedented form of pollution: microplastics.[27]

Disasters became more frequent. When the Brundtland Report was released in 1987, there were 191 disasters worldwide. Between 1999 and 2016, there were more than three hundred per year.[28]

Why did sustainable development fail to prevent these effects while also being one of the most popular concepts in popular culture?

Demographics play a role, of course. From 1987 to 2015, the planet's population grew by 2.4 billion.[29] Increases in births and life expectancy have given more people more time to consume more things. Meanwhile, millions of people in developing countries have become wealthier and have increased their living standards. Much like Ms. Do Santos, millions of urbanites in Latin America, China, India, and other places have escaped poverty. But their consumption patterns have become more similar to those of industrialized nations. They move to larger, air-conditioned homes, travel abroad, own cars, accumulate appliances, replace cell phones, amass clothes, eat more meat, and drink more milk and bottled water.

Some argue that it is time to call for the end of sustainability.[30] Critics believe that there are intrinsic problems with sustainable development. "As with development, the meanings, practices, and policies of sustainable development continue to be informed by colonial thought," argues Subhabrata Bobby Banerjee from University of South Australia. This has resulted "in disempowerment of a majority of the world's populations, especially rural populations in the Third World."[31] "Deep down," says Indian scholar Shiv Visvanathan, "the Brundtland Report still believes that the expert and the World Bank can save the world. All it needs is the application of better technology and management." And then the critic of sustainable development continues, "What it fails to understand is that a club of experts, whether Brandt, Brundtland, or Rome, is an inadequate basis for society. What one needs is not a common future but the future as a commons."[32]

Others reply, however, that it is unfair to accuse sustainability of a collective failure to contain our environmental crimes.[33] Several environmentalists and academics argue that the initial approach described in the Brundtland Report was just a starting point and not an end. They contend that holding sustainability responsible for our sins is unfair because it is a dynamic notion that makes sense only when applied to specific contexts. The idea of sustainability, they claim, must continue to evolve as new challenges are found and our understanding of environmental degradation is updated.[34] The real problem, they argue, is that not enough citizens and decision makers subscribe to the real premises of the green agenda.[35] They point to the lack of practical tools and political will, coupled with idealistic promises but insufficient action from leaders. For many, the problem is not the idea of sustainability per se but the wrong (and limited) manner in which it has been adopted.[36]

Many defenders of sustainable development recognize that sustainability is an imperfect concept. But they contend that it is also our best choice for creating awareness about the dangers we have caused—and will cause—to ecosystems.[37] Criticism of sustainable development is, for them, counterproductive. They fear that it can be used to justify additional disengagement from the environmental movement and its research on solutions to environmental problems.[38] They often feel betrayed by ecologists and environmentalists who choose to attack sustainable development. For them, convincing people to become green is already difficult enough. Applying more radical measures will only discourage the public from adopting good practices. Most seem to prefer saving the baby from being thrown out with the bathwater.

Others do realize the limits of the approach but defend sustainability because they are a part of its economic system. It is obviously difficult to attack sustainable development when you are a part of the green industry as a consultant, researcher, writer, professor, or vendor of green technology. Their reasoning forces them into contentious grammatical gymnastics to try to distinguish between forms of "green," types of sustainability, and interpretations of sustainable development.[39] Some have claimed, for instance, that certain forms of "green" are tools for developing environmental awareness. They say that sustainability is the ultimate goal of living in harmony with the environment, while sustainable development is the means to achieve it.[40] But even these defenders of the idea adopt other interpretations and definitions. To be sure, their sophisticated distinctions are rarely understood, adopted, or shared by consumers, journalists, salespeople, and decision makers.

But many defenders of sustainability have a point. It is true that the green agenda has been adopted rather superficially. There has also been an irresponsible lack of political will to adopt greener forms of transportation, energy production, and urban development. It is accurate to say that sustainable development is a malleable concept. In international policy, for instance, the idea of sustainability has evolved from the work of the Club of Rome in the 1970s to the UN international conferences on the environment (such as the one held in Stockholm in 1972), the Brundtland Commission, the UN conference in Rio in 1992, the Millennium Development Goals fixed in 2000, and the Sustainable Development Goals set in 2015. Still, it is difficult to see how the "malleable concept" argument proves the pertinence of sustainable development today. If anything, it proves only that sustainable development has endured as a key pillar of international policy in environmental protection and disaster risk-reduction.

It is difficult to argue that sustainable development is not one of the most popular ideas offered in international agreements, consulting reports, coffee shop conversations, board meetings, and marketing fairs today. And as these terms continue to gain momentum, it becomes increasingly difficult to distinguish between "green," "sustainability," and "sustainable development." I, for one, have given up on efforts to teach, let alone investigate, these distinctions.

What if sustainable development's popularity is a part of the problem? What if the baby and the bathwater cannot be easily dissociated from each other? In this book, I argue that, most often, the irreducible defenders of sustainability fail to acknowledge its inherent incapacity to produce effective transformation. Sustainable development has surely failed because it has not been properly implemented. But it has also failed because of several factors that are *inherent* to its own premise.[41] I will explore some of them here.

The first one is linked to the meaning of "development" itself. In contemporary economies, development inevitably requires economic growth. And growth only happens where there is consumption.

SUSTAINABLE INCREASES IN CONSUMPTION AND WASTE

As the world population has grown over the past three decades, consumption has (obviously) increased. In many ways though, consumption per capita has also amplified since the incorporation of sustainable development in commercial brochures and coffee table books.

Take housing, for example. In Australia, the average house in 2012 was 38 percent bigger than in 1984.[42] Over the past sixty years, Australian homes have more than doubled in size.[43] From 1973 to 2016, the average single family home in the United States went from 1,660 to 2,640 square feet.[44] In the last three decades, the average house in the Canadian province of Quebec grew by 17 percent.[45] Paradoxically, the average household size in these three countries shrank in the past few decades. As family sizes decreased, Australian, American, and Canadian homes not only became bigger . . . but emptier.

Emptier and cooler. The International Energy Agency estimates that in 1990, there were 0.9 air conditioning units per capita in the United States. This figure rose to 1.16 in 2016. The same report found that Americans consume more energy for cooling than is consumed by the total number of people living in Africa, Latin America, the Middle East, and Asia (excluding China).[46]

Not only did millions of people begin to live in larger and cooler homes, they also started to buy larger, air-conditioned cars. As cars became more efficient due to technological innovation, drivers began to spend more on bigger and more sophisticated vehicles. Jato, a firm that conducts research on the automobile industry, reports that the shift from cars to SUVs, which started in 2011, is still going strong. As sales of SUVs have grown, their share of the market has increased from 22 percent in 2014 to 36 percent in 2018.[47] Americans do not carry all of the blame for this. The same shift from cars to SUVs occurred in Europe. There is no doubt that new SUVs use fuel-saving technology and pollute much less than those built in the 1990s. But SUVs require more material and components, as well as additional energy to be manufactured. As a result, they are still about 30 percent less efficient than smaller cars.[48] Today, transportation accounts for about 14 percent of greenhouse gas emissions, with SUVs increasingly responsible for them.

Traveling by air is another very polluting activity. It generates months-worth of human-generated carbon emissions—and yet, before the COVID-19 pandemic, we were travelling relentlessly. It is estimated that in 2017, there were 2.4 times more air passengers in the world than there were in 2000. This is an increase that obviously outnumbers demographic growth.[49] Perhaps nothing quite captures gratuitous consumption in industrialized countries more than bottled water. But as it turns out, the consumption of bottled water has skyrocketed in the past two decades. By 2017, Americans consumed 2.6 times more bottled water per capita than they did in 1999. And Americans are not alone: per capita increases in the consumption of bottled water (albeit less significant than U.S. increases) can also be found in the UK, France, and other European countries.[50] In the last sixteen years, we have collectively produced more than 3.1 billion tonnes of plastic. The era of sustainable development has also become one of sustained increases in per capita plastic consumption.

As our consumption per capita increased, so did our waste. A study conducted in 2013 found that between 1995 and the late 2000s, there was an increase of 8 percent in per capita municipal waste in countries that are members of the OECD, an economic organization of mostly rich nations.[51] Construction and demolition waste accounted for about 15 to 30 percent of the total waste generated. And a great deal of that waste was linked to our increasing need to renovate spaces and follow fashion trends. Hotels, for instance, are now typically renovated every five to seven years.[52] Makeovers are also common in restaurants, bars, and store brands. Quite often, the reason behind these renovations is the need to keep up with clients' tastes. A similar trend is also found in the residential sector. It is estimated that between 2008 and 2018, there was a 34 percent increase in the sales of home improvement materials and services in the United States.[53] And this is not only a demographics issue: in 2017, the average American homeowner spent over US$1,000 more on home improvements than the year before.[54]

It is true that since the 1980s, people in rich countries have started to recycle much more. But experts estimate that less than 10 percent of plastic consumed in the United States has been recycled.[55] For many years, Americans and Europeans tossed plastic bottles into recycling bins and gave no thought as to where they would end up. But we know exactly where all those plastic bottles ended up—they were shipped to Asia. China and other Asian countries are now refusing to import plastic and paper produced in Europe and North America. The majority of recyclable plastic thus ends up in landfills.[56]

Despite the recent expansion and adoption of more efficient technologies, energy consumption has been on the rise worldwide. The International Energy Agency estimates that global energy consumption in 2018 "increased at nearly twice the average rate of growth since 2010."[57] The demand for more electricity has accounted for over half the growth of energy consumption. One of the main reasons is that 2018 was one of the hottest years on record. That same year also witnessed a very cold winter in many places, including North America. As more people rely on air conditioning and heating, more energy is required. This is terrible news for the climate, because it means that we might be entering a vicious cycle in which we pollute more and heat up the planet by trying to cope with hotter summers, cooler winters, and other climate variations. As many countries developed, global per capita carbon emissions increased by almost 17 percent between 1987 and 2014.[58]

It's simple: we consume more stuff. We have been quicker to adopt the sustainability dialect than to even maintain per capita levels of consumption. Surely, not all of this consumption has a direct effect on disasters. The consumption of plastic, for instance, has only an indirect impact (although we shouldn't forget that waste sometimes contributes to drain blockage, thereby increasing the impacts of floods and water surges). But increased consumption in recent decades exemplifies how sustainability has failed to diminish the factors that fuel global warming and environmental degradation.

We are also consuming more green stuff, including more energy-efficient homes, cars, and machines. Of course, consuming green wouldn't be so bad if many green products didn't belong to companies with skeletons in their closets. But they do.

SUSTAINABLE ELITES AND SUSTAINABLE CONSULTING FEES

"Becoming the leader in sustainable development in the region" and a "world-class city."

These are the goals of Ecopark, a new urban development on the outskirts of Hanoi. When a group of scholars and I visited Ecopark in 2011, the project leaders were proud to show us where the new buildings were going to be built. We rode an air-conditioned shuttle where a smartly-dressed project representative spoke into a microphone, while pointing to the side of the road: "Here we will build the eco-commercial center, inspired by old Vietnamese villages, and here the modern community clinic. The state-of-the-art sports complex will be here, the marina over there, and the tennis courts just near here." The other academics and I had seen the models and drawings of Ecopark before, and the project seemed well-organized. To our surprise, though, the tour guide was now pointing at wetlands, forests, and agricultural land.

Ecopark became a US$6 billion investment and an enormous five hundred–hectare satellite city a few kilometres from Hanoi. With twenty thousand residential units clustered in gated communities, it is one of the largest master-planned developments in North Vietnam. Largely designed by foreign companies, it features urban patterns typically found in modern neighborhoods in Singapore, China, and Taiwan. With curved roads, oscillating pedestrian paths, and abundant vegetation, the inner clusters appear more like modern parks than Vietnamese neighborhoods (see figure 3.3). There are housing solutions for all tastes: residential towers, midrise buildings, townhouses, and villas. There's not much variety when it comes to

FIGURE 3.3 (*Top*) Visitors contemplate the model of Ecopark. (*Bottom*) The 500 hectare project.

Source: Gonzalo Lizarralde, 2011.

clients' incomes, though: the Ecopark developers' targeted client is decidedly Vietnam's booming higher-income sector.

Sustainable development was integrated during the early stages of the project. An innovative illumination system was built in to prevent light pollution.[59] During construction, special measures were taken to reduce stress among workers. This included implementing "biophilic principles" meant to capture the "innate emotional affiliation of humankind to all organisms in nature." A team of researchers who studied the construction site concluded that "happy workers are productive workers."[60] The investors are also pleased. In 2014, Ecopark won the Asia Pacific Property Awards due to its "careful planning and sustainable development strategy."[61] In 2018, it obtained the Best Urban Area Project award at the Vietnam National Property Awards—among other accolades accredited to it over the past few years.[62]

Ecopark promotes a healthy lifestyle whereby residents live "in perfect harmony with nature." The idea is that occupants retreat from Hanoi's noise, pollution, traffic, and the inevitable contact with poverty. Unlike almost every street in Hanoi, there are no street vendors in Ecopark. Guards take good care to ensure that the sidewalks are free of traders and disturbances.

Most residents of Ecopark work outside the satellite city, and therefore, tend to commute to work. For this reason, there are more vehicles than in most developments around Hanoi.[63] Work aside, they have little reason to leave Ecopark. There are brand-new schools, lakes, playground areas, gyms, and swimming pools, in addition to a pharmacy, a private university, a shopping mall, and ample parks and manicured gardens. Want to attend a music or arts festival? Play golf? Go out to dinner or for a coffee? There's no need to set foot outside Ecopark.

At first glance, Ecopark is a seemingly idyllic setting. Children run freely and safely, and adults ride bicycles and enjoy local cafés. Its residents contemplate flower gardens and meditate near the lakes, and they all enjoy year-round cultural and artistic events. While everything in Ecopark appears to revolve around sustainability, a closer look behind Ecopark's leafy trees and blooming flower gardens reveals one of the most recent stories of humiliation, poverty, and violence in Vietnam.

When my colleagues and I noted that Ecopark was being built on agricultural land and wetlands, we were not wrong. It is estimated that as many as 2,900 families were evicted or expropriated for the construction of Ecopark. Many of them were peasants coerced against their will to give up their land.[64] Former farmers were forced into unemployment, losing their livelihoods and, by extension, falling into debt. They lamented that the compensation they received for their land was equivalent to "no more than four to five months of revenue derived from their prosperous farms."[65] One of them explained, "People didn't want to sell the land because farmers have to have land, just like factory workers need factories." He wondered, "Now that we've lost the land, what should we do?"[66]

Villagers and peasants affected by Ecopark organized a resistance movement. In 2006, construction in Ecopark was temporarily suspended amid protests by residents who had lost their properties. Protests erupted again in 2009 and 2012: hundreds of villagers camped out on the land overnight and organized demonstrations. A force of more than two thousand police officers and unidentified men without uniforms met the protesters.[67] The men, armed with stones and Molotov cocktails, fired tear gas into the crowd of citizens. Several farmers were arrested, and community leaders and journalists were intimidated. Other forms of humiliation followed. Local cemeteries were desecrated, and even elderly women and mothers of war veterans were beaten.

These demonstrations became one of the most important social conflicts over land in the country.[68] The government tried—but ultimately failed—to prevent the circulation of news about these events. Months later, residents submitted a complaint to the Van Giang People's Court, suing the district chairman over the evictions. But the court returned their case, arguing that "there was not enough evidence" to file a complaint.[69]

All this violence and social injustice was enough to make the sustainable approach of Ecopark sound absurd. But the farce did not end there. In Ecopark, sustainability was only a luxury for the wealthy Vietnamese elites; a cause and consequence of the gap that extends between them and the poor. The Environmental Justice Atlas, a database of social injustices edited by the Universitat Autònoma de Barcelona, argues that the perception put forward by the urban Vietnamese upper classes "drastically clashes against the needs and development visions" of the evicted families.[70] Researchers find that Ecopark and other similar "green developments" in Vietnam emulate a type of market modernism based on "allusions to urban forms and images from other successful cities." This type of master-planned, profit-seeking, green development reflects the elite's "aspirations to emulate a global lifestyle" and to belong "to the global world of success, wealth and privilege."[71]

Lisa Drummond, an urban studies professor at York University in Canada, has studied Hanoi for decades. She states that a "chasm has begun to open up" between rich and poor in the city. Developments such as Ecopark reflect—and also perpetuate—those inequalities.[72]

Ecopark now also produces plenty of pollution. Much like tourist resorts, the gardens in Ecopark are sprayed with pesticides that kill insects, frogs, and snakes. So it seems that residents are pleased with the grass, trees, and flowers. But they're less than thrilled with the fauna that comes with Ecopark's "natural" landscape.

There haven't yet been any major floods directly attributed to Ecopark yet. Floods, however, are likely to occur in the traditional villages surrounding the development due to its design. Researchers and residents know that peripheral developments in Hanoi are typically built on agricultural and unoccupied land that plays a fundamental role in absorbing rainwater and regulating the flow of rivers and streams.[73] New developments on this land contribute to the recent increases in the frequency and intensity of floods and water surges in traditional villages.[74]

How does one explain that a project that has caused so much suffering—and that may provoke further social injustices in the future—is considered an example of sustainable development deserving international awards?

The answer lies in the proximity between real-estate investors, urban developers, and politicians, says Danielle Labbé, a Canadian professor who is an expert in urban projects in Vietnam. Danielle told me that political and economic elites regularly collude to create new developments on the peripheries of Vietnamese cities. This often involves expropriation of land and unfair compensation for displaced farmers.[75] Vietnam's city-making business is based on "interlocking combinations of entrepreneurs and government units." These coalitions "control property, convert land from lower to higher value categories, and intensify future land uses on the rapidly urbanizing outskirts of cities."[76]

Is sustainability an accident in this real-state, profit-seeking scheme? Hardly. If anything, Ecopark exemplifies how sustainability has been hijacked by companies and politicians to amass money and power. It is also an example of the way that sustainability is appropriated as a marketing strategy, regardless of the values conveyed in project plans. In Ecopark, and in an increasing number of similar endeavors in the city-making business, sustainability has been voided of its moral worth.

SUSTAINABLE DEVELOPERS AND COALITIONS

The relationship between political and economic elites predicated on the idea of sustainability is not at all rare. This proximity has resided at the core of sustainable development since its official, international introduction in the Brundtland Report.

Already in 1987, the report argued that cooperation between governments and industry was crucial for more sustainable forms of development. Noticing that private and public land is quite often available in cities, the report observed that "several countries have introduced special programs to encourage public and private cooperation in the development of such lands, a trend that should be encouraged." In addition, the report argued that many countries have managed to reverse inner-city decay and its associated economic decline through "cooperation between the public and private sectors, and significant investments in personnel, institutions, and technological innovation." In regard to research ventures, the report recommended that specific responsibilities should "be assigned to institutions and corporations." This type of agreement, the commission asserted, would allow "for the equitable sharing and widespread diffusion of technologies developed."

Capital from corporations and investors is necessary to achieve growth in poor countries. The Brundtland Commission claimed that "if large parts of the developing world are to avert economic, social, and environmental catastrophes, it is essential that global economic growth be revitalized." This was to be achieved through "free market access for the products of developing countries, lower interest rates, greater technology transfer, and significantly larger capital flows."

The recipe used in Ecopark was drafted more than three decades ago in the founding documents of sustainable development. By constructing the idea of environmental protection around a coalition of political and economic power, sustainable development paved the way for the types of abuses we find in Ecopark and in many other urban developments today. Sustainability has harvested a fertile ground for insincere marketing strategies and political games.

Unsurprisingly, many have found that sustainable development sits comfortably in the spectrum of neoliberal policies (the Vietnamese case also suggests that the model does not discriminate between capitalist democracies and communist governments). Fervent critics of sustainable development and resilience include Brad Evans, a UK political philosopher and professor, and Julian Reid, a professor at the University of Lapland, Finland. In their book, *Resilient Life*, they contend that there has always been "a strategically manipulable relation between the two doctrines" of sustainability and neoliberalism.[77] For them, sustainability has been the perfect glue with which to bind economic and political interests under the guise of protecting nature.

But a coalition of financial and political power to optimize the benefits of a few cannot thrive for a long time without public support. Here is where the "common future" argument, popularized by sustainable development discourse, comes in handy.

OUR NOT-SO-COMMON FUTURE

A shared narrative in sustainable development is that we are all responsible for "saving" the planet. Surely, creating general awareness of our role in environmental degradation is crucial today. Climate change, pollution, and an increasing number of disasters are caused by our collective action. We all contribute—directly or indirectly—to displacement and unnatural disasters on local and global scales. But do we all really bear an equal share of responsibility for this mess?

A 2016 article published in the prestigious journal, *Science*, suggests that we do not.[78] According to the study presented in the article, ninety companies are to blame for most of the world's climate change. The paper shows that nearly two-thirds of anthropogenic carbon emissions derive from a limited number of companies and government-run industries, including Saudi Aramco, Chevron, Gazprom, Exxon Mobil, British Petroleum, and other oil companies.

When the study was released, it sparked a heated debate. Many argued that, in order to achieve progress, it is important to have a target in the fight against climate change. They found that pointing to specific industries or political players is a vital first step to starting legal actions against the world's worst polluters. Others were less convinced, arguing that it is both unfair and misleading to blame oil companies for the choices we make as citizens on a daily basis. They contended that oil companies simply respond to the demand of consumers, who enact their free will in their consumption choices.

Richard Heede, cofounder of the Climate Accountability Institute and the author of the study, reminds everyone that consumption is not enacted in a vacuum. "I, as a consumer, bear some responsibility for my own car," he says. But "we're living an illusion if we think we're making choices." He rejects the idea of a total free will in consumption patterns "because the infrastructure pretty much makes those choices for us." He then points to a clear example, reminding us that during the 1980s, "the vast bulk of federal energy subsidies went to conventional energy sources." It is difficult to argue that citizens have a natural taste for driving more, flying more, and burning fossil fuels. It is easier to accept that their choices are framed by the "solutions" presented to them. According to Heede, a great deal of responsibility must be attributed to government support of corporations and polluting companies.

Multinational corporations and political power aside, it is likely that we do not share the same responsibility as *individuals*, either. A study published by Oxfam in 2005 found that the richest 10 percent of people in the world are responsible for around half of all global emissions. The same study showed that the poorest half of the global population is responsible for only around 10 percent of global emissions. Ironically, these low-income people live in the countries most vulnerable to climate change.[79] (Remember the case of Cuba in chapter 1.)

Sustainability rarely benefits all in an equal manner. Ecopark, for example, is not the only project in which sustainability only benefits the rich. Urban interventions developed under the sustainability mantra are used to embellish the wealthiest districts in cities across Europe, North America, and many other areas. In some cases, green infrastructure projects, parks, and landscapes aimed at attaining more sustainable cities have been a significant source of gentrification. These projects often push low-income families to poor areas and displace them from their original locations. Such green gentrification processes have been recently reported in New York, London, Montreal, and Barcelona, among other cities.[80]

This is where a major part of the problem lies today. Sustainable development has always served a partnership between political and economic elites in the forms of subsidies, partnerships, expensive consultants, special concessions, and ownership. It has given us some of the ninety most polluting entities responsible for the majority of climate change. It is precisely this coalition of power that has laid the infrastructure for current conditions and has determined many of our seemingly free choices as consumers—from where we buy our home, to what type of vehicle we drive, and what sort of food we eat. As citizens, we have inadvertently relied on a coalition of financial advisors, corporate boards, urban consultants, and politicians to solve our environmental challenges. But this coalition can't fix our mounting problems, as it actually produces and further fuels most of them.

In this context, the narrative of shared concerns, challenges, and endeavors promised by Brundtland and other sustainable developers takes a different perspective. It becomes clear that we are not all in the same boat. The collective agreement put forth by sustainability has, again, proven to be an illusion.

FIFTY SHADES OF GREEN

My friend who lives in a Canadian suburb is vegan but drives an old Honda Civic to work. My colleague eats lots of meat but lives in an apartment building in Toronto and rides a bicycle to her office. My cousin, who is both a cyclist *and* vegan has moved to a house in the countryside, where he capitalizes on the forest each day. Which one is more environmentally responsible? Which one has adopted the greenest approach?

These are questions that are almost impossible to answer without a comprehensive life-cycle analysis of carbon footprints, pollution, and other impacts. It would take a sophisticated "eco-meter" to gauge individual environmental virtues.

Eco-meters do not actually exist, but in 2012, my colleague Benjamin Herazo, an expert in sustainable development, and I tried to improvise one. We wanted to study a green urban project developed by our own university in Canada. The project we wanted to investigate had been originally sold as a green campus. We wanted to know where the client, professionals, constructors, and other stakeholders stood in terms of sustainability. We set up a scale that ranged from the most radical transformers (i.e., those who advocate for substantial changes to protect nature) to the more status-quo stakeholders (i.e., those who are comfortable with conventional solutions or minimal changes to attain sustainability). Hundreds of documents and dozens of interviews allowed us to place project participants on our eco-scale. Benjamin then followed their sustainability approaches for five years. The results obtained in 2018 were useful for understanding how sustainability can create a deceitful feeling of collective agreement.[81]

In the early stages of this green district project, stakeholders had adopted different approaches to sustainability. There were, of course, radical environmental activists, conservative conformists, and many people in between. The choice of location for the project had been very controversial (it sits between one of the richest and one of the poorest neighborhoods in Canada). The client, therefore, needed public support for this new urban development. Neighbors and regular citizens were therefore invited to come up with new ideas to create a model neighborhood.

When these ideas began to be drafted, a sort of collective enthusiasm emerged. The client, the city, designers, consultants, and neighbors began to imagine a district full of green roofs, solar panels, sophisticated systems for water and waste management, bicycle paths, lots of green areas, vegetable gardens, and, of course, hundreds of trees. Drawings of the project in 3D became (literally and metaphorically) greener. Participants started to dream about the most ambitious green neighborhood certifications. Even the more conservative participants began to advocate for ambitious sustainable measures. Almost all of them appeared to be, for a moment, environmentalists engaged in a "common future."

It was then time to develop more precise plans, to review compliance with regulations, and estimate costs. A dramatic reality check settled in. Green roofs, solar panels, vegetable gardens, and other green gadgets suddenly disappeared. Water and waste management systems were once again conventional. Bicycle paths and green areas shrank. And not as many trees decorated the new 3D drawings. The green enthusiasm and consensus quickly vanished.

Months later, Benjamin and I confirmed that conventional buildings were built. The client, the city, consultants, and other stakeholders settled for a traditional district rather than a green campus. The green solutions gave way to squared boxes with glass façades and opaque sides of grey brick. A few environmentalists kept pushing for the greener version that had been codesigned in the early, enthusiastic meetings, but the client argued that the budget was not enough for them to realize that ambitious vision. Those opposed were no longer invited to meetings, and a marketing company was hired to sell the benefits of the new project and garner public support.

It might be surprising to witness the ease with which a deeply engaging green initiative became careless and conventionally grey. But what happened is not at all rare. Even my architecture and urban planning colleagues can't agree on the price to pay to protect the environment. In believing that we've always had a common problem, approach, and solution, the concept of sustainability has clouded the different actions we are willing to take to protect nature. This false sense of collective agreement has limited the discussions and social debates most needed today. It has also opened the door to hollow, green marketing strategies and consultants' reports. Now, almost anything can be sold as "sustainable."

This has had dramatic consequences in our cities. We find it difficult to distinguish between the real measures that protect the environment and greenwashing marketing schemes that have perpetuated common myths. For millions of Canadians and Americans, suburban life with its manicured lawns, plastic swimming pools, and views of agricultural landscapes, is still viewed as an environmentally conscious lifestyle. Living in an apartment building in a dense city center, with its proximity to traffic, smog, and few lawns, is, for many, the antithesis of a sustainable life. And yet, nothing could be further from reality. As we try to "connect" with nature by moving to remote areas, we increase commuting, urbanize land, canalize rivers, pollute water sources, and build new infrastructure. Our green attempts to reconnect with nature are also often accompanied by pesticides and animal traps. As it turns out, we love a selected number of flora species—but not others, which suburbanites perversely call "weeds." And we prefer not to be in contact with most of nature's fauna, including insects, certain birds, and rodents.

In places at risk of disaster, sustainability has been manipulated to achieve the notion of development pursued by the most powerful. This has distracted attention and resources that are needed to achieve effective positive transformation. In some cases, this has also perpetuated social injustices and created new conflicts.

The strategies we have seen in this chapter are carefully calculated and rely on preserved ignorance to succeed. At the same time, the vast majority of people, charities, and organizations I have come across have adopted sustainable development with *good* intentions. For them, sustainability has never been a perverse scheme of disaster creation but a virtuous act of disaster prevention. We will now see what these well-intentioned, virtuous people attempt to do in the name of sustainability. We will also observe how difficult it is to protect people and the environment through sustainable development—even when acting in good faith.

THE GREENEST BUILDING THAT CAN NEVER BE

"To create the greenest office building ever made in Canada"—that is how Norbert Reims explained the objective to me in 2002.[82] A few years before, Norbert's team had created an environmentalist NGO that quickly became an international benchmark for sustainable development. As the organization grew in size and importance, Norbert and his team realized that they had to build a project that would provide an indisputable example of good practices in green construction. The Sustainability Institute would become a permanent hub of knowledge production, training, information sharing, activism, and consulting in areas linked to environmental protection and justice. The building had to showcase best practices in design, construction, and operation. It also needed to comply with the highest standards in materials and energy reduction, as well as users' health, comfort, and well-being.

Norbert and his team are committed and disciplined. I always found them to be the type of people who consistently set the example for a no-waste, low-carbon lifestyle (when I signed a research contract with Norbert, he pointed out that printing a paper copy of it was unnecessary, so I hid my hard copy under my laptop). When it came to designing the new building, Norbert's team hired the best architects and worked closely with the design team to produce innovative solutions. They also conducted intensive research to find less polluting materials and the most efficient systems. They made efforts to hire the best constructors, and additionally, they documented each stage of the design, construction, and operation phases. Everything was in place to obtain the highest LEED certification. Perhaps most importantly, Norbert and his team treated their building as a laboratory to test the most radical changes in green construction.

In 2011, after many years of devoted work, the Sustainability Institute opened in downtown Montreal. The building hosts a number of charities and organizations committed to sustainability principles and social causes. As such, the institute became a magnet for environmental activism and research. But Norbert and his team did not stop there: they wanted to know how successful (or unsuccessful) the building was. They decided to do what very few owners dare to do: to evaluate the performance of their own building and make the results public. Their goal was to contribute to public knowledge through total transparency, even if the results could expose their own mistakes. They asked my research team to help with this assessment.

Norbert's courage paid off. Thanks to his hard work, we now know how hard it is to achieve sustainability in contemporary office buildings. While Norbert and his team had assumed that they could reduce energy use to 30 percent of what a regular office building might consume, those ambitious reductions never happened. As a matter of fact, the study proved that their building consumed more than twice the energy that they had expected. To be fair, the Sustainability Institute still performs much better than most standard buildings[83]—but the results still show the glass half-empty for Norbert. "Most conventional buildings are made to save or make money, not ecosystems," he said to me. "They have embarrassing carbon footprints. This can't be our benchmark."

In our contemporary context, every bit of carbon not generated is badly needed. But the Sustainability Institute case teaches us that achieving significant reduction is very difficult. It is a tricky task even for the most committed, full-time, environmental activists—let alone the average profit-seeking developers.

The underperformance of Norbert's building might be surprising at first—yet a similar pattern has been found in countless "green" buildings across the United States, the UK, Europe, and other places. A study conducted in 2016, for instance, found that ninety-one LEED-certified buildings in New York City "are not necessarily more energy efficient than typical buildings."[84]

There are many reasons for this underperformance. Building users do not know how to maintain and calibrate high-tech equipment. Estimations are often based on people's expected behaviors and not on their actions. There is still a lack of data, tools, and methods with which to estimate and analyze carbon footprints. Additionally, certifications constantly evolve, making it difficult to compare standards at different moments in time. But fixing these problems is within reach and might actually constitute the easiest part of the challenge. The most difficult part is linked to the principles of design and the characteristics of modern construction. So, let us assume that we are now able to reduce the impacts of lousy building operations, wrong estimations, and the lack of data, tools, and methods: How can we design the optimal sustainable office building today?

Well, as Norbert and I eventually learned, we can't—at least not by reasonable means.[85]

One reason is that performances in buildings or infrastructure are not cumulative. For the past fifty years, office and commercial buildings have increasingly relied on machines to bring in fresh air; to cool it, filter it, and distribute it homogenously. Machines are also required to extract stale air from rooms and evacuate it. Mechanical and electronic systems are now used to heat floors, move people up and down, control lighting, pump water, program fire-protection systems, control appliances, as well as conduct a series of tasks that allow buildings to offer a certain level of comfort. Performances can be obtained in each of these systems to reduce energy consumption. Once performances accumulate, though, the systems start having negative effects on one another. For example, machines used to distribute cool air produce heat, which in turn, require stronger machines to cool the air. This limits the possibility of improving the overall system. At a certain point, the performance of these machines reaches a level after which improvements are modest and very expensive.

A second reason is that green buildings are often more costly than conventional ones. Experts estimate that a green-certified building can be 15 to 30 percent more expensive, which is why they are then sold or rented at a premium rate. A recent study in London found that green certifications in the office market results in a premium of about 20 percent for rents and almost 15 percent for sales, relative to noncertified buildings in the same areas.[86] This means that green buildings tend to be commissioned or acquired by clients who have the means to afford those extra costs; in other words, richer clients who have high expectations in terms of comfort and finishes. As one mechanical engineer recently told me, "A green building is like a Ferarri." He went on to explain, "Wealthy business people won't pay thousands for an uncomfortable Ferrari, or millions for an office space that is poorly lighted, or where the inner temperature is too hot or too cold."

Here lies the conundrum: as buildings rely increasingly on technology, improvements in users' comfort require more complex machines and systems. And there are rising expectations for indoor conditions, such as temperature, quality of air, ventilation, lighting, and sound control, among others. In turn, these expectations push designers and builders to create sophisticated and machine-dependent buildings. This triggers a vicious cycle in which technology leads to mechanized solutions for comfort that then also require increased mechanization. The overall result is that compliance to this demand for comfort is now offsetting part of the energy reduction that green buildings can offer.

The third reason is that a great part of our energy consumption derives from the relationships between buildings and *not* the way individual buildings are made. It is the distance between buildings that generates the need for transportation, additional heating, and a multiplication of building systems. "Overachiever" buildings that are disconnected and work independently do very little to reduce these forms of energy consumption. Because developers and professionals work on individual initiatives one at a time, they are (stubbornly) focused on the performance of individual buildings and not on overall urban-system efficiency. Revealing some of the limits of current approaches to green development, an article published in *Nature News* concluded that urban policy usually addresses ecological issues on the scale of projects or buildings, thus failing to understand the complexity of current urban systems.[87]

The fourth reason concerns a common lack of focus on embedded energy. Significant efforts have been made to reduce energy consumption in buildings and infrastructure. But little is still done to reduce the energy required to *execute* them; that is, the building's embedded energy.[88] Other reasons are connected to deficiencies in current carbon reduction measures and the lack of energy reduction standards. When those standards exist, they are often based on anticipated performances during the design process, not on effective ones during operation. In many countries, standards for construction are based on minimum codes (such as minimum areas and dimensions)—not maximum ones. Clients are thus entitled to oversize buildings, parking spaces, and homes without efficient restrictions. Finally, it is fair to say that the principle of sustainable construction in most countries has been to establish standards that represent no harm to the market.

As we shall now see, even well-intentioned clients can have limited capacity for grasping the complex and dynamic nature of environmental damage. Unfortunately, the tools and methods used for sustainability do not help them much in this regard.

DECEIVING GREEN

At first, Laura Lachapelle thought she'd received the mandate that every environmentally-conscious architect would envy: her client wanted to renovate his house to comply with the highest standards for reducing energy consumption.[89] And the client was willing to put his money where his mouth was. He wanted the greenest home possible.

As the main architect of this project, Laura suggested building the home according to the Passive House certification—one of the most demanding labels that exist today to certify buildings that are built to reduce energy consumption.

Laura's client accepted and made this his priority. One of the key aspects of the Passive House certification is that it sets up a maximum energy use per square meter of built area. The aim is for owners and designers to reduce, as much as possible, the energy used to heat spaces in winter and cool them in summer. To achieve this objective in Nordic countries, layers and layers of insulation are necessary, making exterior walls very thick. Windows often have three layers of glass. Roofs are overdesigned to prevent the transfer of temperature between indoors and the outside.

These features bear an obvious disadvantage: significant amounts of energy to produce, transport, and install the extra material for the appropriate levels of insulation. Nonetheless, Passive House is one of the most popular certifications worldwide, as it encourages drastic reductions in the energy used by mechanical systems to obtain indoor comfort.

Like most architects who work on sustainable homes, Laura used computerized models during the design phase. Through these models, she simulated the building's energy consumption. In early simulations, Laura's design achieved a very good performance, though it was a bit higher than the threshold required to obtain the Passive House certification. In order to receive the certification, the house had to have fewer and smaller windows (even the best-performing windows are poor insulators against temperature). Laura's client refused this solution, however: this was an expensive home, and expectations were high. Adequate natural lighting and expansive backyard views were nonnegotiable.

The client, instead, proposed an alternative solution: increase the size of the house. The client realized that an almost constant amount of energy consumption is required for *any* space regardless of its size. He also learned that, in his desired certification, consumption is measured as the ratio from energy used to the area of construction. So building a bigger house implied that the amount of energy used would be divided by a larger number of square meters of construction. In this way, the project would achieve the threshold required by the certification.[90]

The architect tried to convince her client that this solution did not make sense from an environmental viewpoint. It was a correct mathematical exercise but a deceiving measure for reducing carbon footprints. A bigger home would require more material and energy to build, offsetting the benefits gained in expensive insulation and sophisticated materials. Yet, by this time, the client wanted a Passive House and would not reverse his decision. A bigger house was eventually built.

Laura's client later moved to his certified Passive House. But the home produced carbon emissions and used additional materials for no real purpose other than obtaining a mere piece of paper—a certificate that could be framed and hung on a wall (as it turned out, a very thick and expensive wall).

When I met Laura, she knew that she was not the only one facing this problem. Increasing numbers of clients and professionals, enchanted by sustainability, are producing solutions that improve their self-esteem and reduce their guilt but have little environmental value. Sustainable development amplifies this tendency.

Sustainability requires concrete ways of assessing change. Industry and governments have, therefore, created hundreds of green certifications aimed at endorsing what they consider "good practices." But these certifications have also become omnipresent marketing tools. In the rush to adopt labels, we have confused the means with the objectives. We have focused our attention on symbolic changes that create a buzz, although they do not necessarily reduce the impact of our unrestrained consumption. As customers, we have deceived ourselves by accepting solutions without seriously questioning their environmental and social value. As David Owen, an American journalist and writer, said, "The point isn't that we're all hypocrites, although of course we are. . . . The point is that, even when we act with what we believe to be the best of our intentions, our efforts are often at cross-purposes with our goals."[91]

Certainly, there are millions of consultants and companies that have little regard for nature. Most of us, however, do not act with malice. Some of the decision makers in Rio de Janeiro perhaps did want to halt deforestation. Most residents who move to Ecopark likely do have an honest desire to connect with nature. And urban developers and designers, such as the ones we have met here, typically want their buildings and solutions to produce a positive impact. But by adopting sustainability, they endorse dubious decisions that sometimes destroy ecosystems and exacerbate social injustices.

IT IS TOO LATE TO BECOME GREEN

The underlying premise of sustainability is that we can maintain our lifestyle while making minimum sacrifices to save nature. Ever since sustainability became a thing, our appetite for consumption has only grown. Sustainability has become a practical—but also deceiving—way of coping with our collective and individual guilt for destroying nature.

We do this in business activities as much as in our leisure and domestic ones. Our love for nature often leads us to take actions that negatively impact it. Our travels to enjoy beaches, forests, mountains, lakes, rivers, and remote islands produce waste and gas emissions, while necessitating the construction of infrastructure. Surely we can talk ourselves out of our guilt by arguing that our latest trip to the Caribbean or Africa generated badly needed jobs. We can also convince ourselves that our participation in last year's conference or industrial fair made the world better. The reality, though, is that we have individual goals that we want to achieve, and we forget (or ignore) that we are placing nature at risk in the process.

When we realize the impact of our actions, we rely on technology to solve the very problems we have created.[92] (I will further expand on this subject in chapter 4). Sustainable development has fulfilled our needs because it is intentionally ill-defined. It is an umbrella with which to cover a wide range of approaches to environmental protection. It allows us to pretend that a broad consensus exists. So even when we have widely adopted sustainability, we still haven't agreed on what it means or how to achieve it. The contradictions are increasingly visible. Sustainable development requires a close alliance between governments and companies. Thus, we have endorsed a fertile ground with which to facilitate cheap marketing and political games.

Sustainability has been hijacked by companies to sell more and by politicians to earn votes.

In recognizing these contradictions, John Livingston, a celebrated Canadian environmentalist, called sustainable development a "full-blown oxymoron." He said that sustainable development, maximum sustained yield, and resource conservation are nothing more than "attempts to sugar-coat our ongoing intention to continue to exploit nature for our own, often indefensible, ends."[93] As we will see in the next chapter, resilience has also become a way to sugar-coat our intention to produce the type of change that benefits only the most powerful.

4 Resilience

"They Say That We Must Adapt"

Incredibly, there was talk of favoritism, of corruption. With its customary discretion, the Company did not reply directly; Instead, it scrawled its brief argument in the rubble of a mask factory. This apologia is now numbered among the Sacred Scriptures.

—Jorge Luis Borges. "The Lottery in Babylon"

THEY BEND BUT THEY DON'T BREAK

"Nou se wozo; nou pliye nou pa kase," quotes the report in the original Haitian Creole. The translation of the proverb follows: "We are like a reed; we bend, but we do not break." The quote and translation are included in a report published in 2017 by the World Bank. The document claims that Haiti should move "from reconstruction to resilient urban planning for a bright future."[1] A good idea, you may think.

But like many other previous reports on Haiti, this one states that the lack of urban planning, regulations, and studies account for many of the country's problems. Like many other reports on Haiti, this one fails to recognize that there have been hundreds of urban plans, regulations, and studies in Haiti over the past twenty years. In 2017, my research colleagues and I found that there were over two hundred studies centered on urban issues in Port-au-Prince alone. In fact, there have been more than ten policy documents related to urban issues adopted since 2010 and more than thirty-seven urban plans for the capital city.

Like many other previous reports on Haiti, this one fails to acknowledge that the problems of Port-au-Prince did not stem from a lack of planning but a lack of action. There are ample policy documents, regulations, laws, and urban plans in Haiti, but the real problem is that these laws are not enforced; policies are not implemented, and plans are not built.

Like many other reports on Haiti, the World Bank presents a detailed analysis of housing and infrastructure deficits, a diagnostic review of poor local governance, a prescription of how problems will increase with more urbanization and demographic growth, and of course, a few sentences on the resilience of the Haitian people.

According to the report, the Creole proverb captures Haiti's "long history of resilience in the face of slavery, colonialism, political oppression, widespread destruction from natural hazards, social exclusion, inequality, and poverty." This history, the authors affirm, "determines its current challenges to development, but most importantly, the opportunities that lie ahead."

That same year, on the opposite end of the world, Prime Minister Narendra Modi of India tweeted: "The people of Gujarat are blessed with a strong spirit of resilience. These floods will not impact the development journey of Gujarat."[2] Modi wrote this after flying over houses and farms affected by torrential rains. The Gujarat floods had killed some 213 people and forced the evacuation or relocation of as many as 130,000. Vast areas of agricultural production were destroyed.[3]

Journalists reported Modi's tweet, but few noted its peculiarity. This wasn't a trivial message. It confirmed that resilience jargon had taken hold in Gujarat, just as it had in Haiti a few years before. This tweet confirmed the tendency to use resilience as a vehicle for the country's "development journey"—a trend that can be traced from Port-au-Prince to Mumbai, and from New York to Cape Town.

THE DISASTERS THAT NEVER OCCURRED

Before writing *Unnatural Disasters*, I thought about writing a book about the disasters that didn't happen. I imagined a first chapter devoted to the Y2K—the much-anticipated technological cataclysm that didn't occur during the first few hours of the new millennium. The second chapter would perhaps cover the 2003 SARS epidemic, another predicted tragedy that humanity largely avoided. I would explain how in 2016, Paris was still relatively functional when the Seine River flooded into some areas of the French capital. Another section would recall that, in 2017, a magnitude 7.1 earthquake sturdily shook more than twenty-one million people in Mexico City, but it caused fewer than four hundred deaths. Other sections of the book would demonstrate how quickly people recover from the disasters that *do* take place. Several examples would be used to illustrate this point: Hurricane Harvey caused major destruction in Texas in 2017, but significant parts of the city functioned normally to avoid additional distress and fatalities. In less than twenty years, New York City has recovered from several major catastrophic events, such as the 2001 terrorist attacks, a massive blackout in 2003, the financial crisis of 2008, super storm Sandy in 2012, and the 2020 COVID-19 pandemic. And yet, New York is still one of the most powerful and vibrant cities in the world. Toward the end of the book, I would praise civilization for overcoming a period in history where urban fires, lighting strokes, diseases, and car accidents are major causes of mortality.

I even thought of a title for such a book: *The Resilient Societies*. I suspect it would have been a commercially successful publication. We all appreciate stories about people and communities that overcome adversity, fight hardships, and win against all odds. And disaster resilience is the ultimate uplifting story (the book would have probably been shelved in the popular personal development and well-being section).

By the end of this chapter, I hope you will understand why I didn't write *The Resilient Societies*. But I must clarify that it is not because I fail to recognize the human capacity to solve major problems and face adversity. We should certainly celebrate that contemporary societies have managed to reduce the deaths and injuries caused by several natural and human-caused hazards. In chapter 7, I will stress how we can all learn from these experiences to further reduce disaster impacts today.

Resilience is now one of the most popular topics in urban consulting and policy, corporate strategic planning, and international agreements. Thousands of books and courses are sold every year on how to make you, your marriage, and your business more resilient. World leaders and politicians celebrate how resilient their citizens are. Thousands of social media networks and groups are devoted to resilience action, and annual conferences are held on resilient housing, agriculture, infrastructure, services, and pretty much everything else you can think of. Policy is made and modified every day to achieve resilience in transportation, healthcare, and commerce. "Building resilient persons, communities, and institutions has become the *sine qua non* of contemporary forms of liberalism," writes Michael Watts, a professor at the University of California, Berkeley.[4] Resilience is a multimillion dollar industry in the consulting sector. It is promoted by multinational organizations such as the Rockefeller Foundation, the C40 Cities Climate Leadership Group, ICLEI, the United Nations Habitat, the United Nations Development Program (UNDP), and the United Nations Office for the Coordination of Humanitarian Affairs, among other giants.

Resilience is increasingly viewed as a win-win strategy for dealing with urban challenges. And since it has now become a central idea in urban consulting services, the tone of its messaging is more prescriptive than ever: Businesses can only thrive *if* they embrace resilience. Individuals ought to develop resilience to avoid suffering. Communities and cities *must* become resilient in times of climate change.

The doctrine of resilience should not merely be accepted by individuals and communities but by territories, countries, the world. It is no longer a way of explaining how we deal with adversity; it is a norm that we must comply with. Today, resilience is predominantly used as a global and corporate term—a mode of speaking that transcends culture, traditions, and language. In 2015, *New York Times Magazine* argued that "almost any organization you can think of has squeezed 'resilience' into its mission statement."[5] While this claim might be exaggerated, it is certainly true of charities, consulting firms, and international agencies interested in development and disaster response. Resilience is so trendy that even Mr. Modi in India, and the World Bank in Haiti, seem to overlook our (very recent) past *before* resilience.[6]

THE WORLD BEFORE RESILIENCE

In 2011, I was a guest of Narendra Modi's government and was invited to participate in a conference organized by his administration in Gandhinagar. At the time, Modi was not yet one of the most influential leaders in the world, though he was already a powerful man. He had been chief minister of Gujarat for about a decade, and I could tell that he was set to amass more power.

One hundred attendees—mainly representatives of the Indian government and journalists—gathered in a modest conference facility. There were also a handful of local and foreign scholars. Bodyguards entered before top members of the Gujarat government made their way into the room. When Modi emerged from a crew of guards, everyone stood up. He slowly moved toward a podium decorated with hundreds of flowers—so many that it appeared as though Modi addressed the audience from a flower garden. His speech was short and, as I later understood (after translation), powerful.

We had been summoned to commemorate the tenth anniversary of the Gujarat earthquake. The 2001 quake had killed almost fourteen thousand people and injured another 167,000. It also flattened over two hundred thousand homes, three hundred hospitals, and 450 villages, leaving 1.7 million people homeless.

After the disaster, an ambitious reconstruction process was put in place. One of the first measures by Modi's government was to create the Gujarat State Disaster Management Authority. This new institution was in charge of managing resources obtained from the World Bank and the Asian Development Bank, along with their government budget. One of the key players in the authority was Venkatachalam Thiruppugazh, a soft-spoken man who was completing graduate studies in disaster management. He also conceived the main principles for the reconstruction efforts. Thiruppugazh came across to me as an intelligent officer who was capable of elevating decisions to an official status. He chose his words carefully and moved with ease among both scholars and government technocrats.

With the help of Thiruppugazh, the new institution made a bold decision. He and his team wanted to avoid the traditional strategy of providing finished homes for disaster victims as much as possible. He knew that quite often disaster victims dislike the residential projects built by authorities and charities. So his team let NGOs and contractors join in under a specific partnership mechanism but put most of its efforts into providing cash subsidies to hundreds of thousands of disaster victims. The authority was convinced that by giving cash resources to affected Indians, they would be better able to respond to their own needs and expectations. Households would also take responsibility for their initiatives, prompting them to use resources wisely and focus on their top priorities. This approach, known as owner-driven reconstruction, had never been tested in India at such a scale. But given the amount of buildings and people impacted by the earthquake, Thiruppugazh knew it was his best choice.

The strategy was successful. There was much to celebrate at the meeting in Gandhinagar, ten years after the disaster. Almost two hundred thousand homes, and thousands of schools and facilities, had been reconstructed in Gujarat. Beneficiaries who had managed their own resources and reconstruction efforts were pleased. These families were far more satisfied than those who had received completed houses built by contractors or charities. Thiruppugazh's bet had paid off.

The main reason I had been invited to Gandhinagar was to explain why and how owner-driven reconstruction processes had been so successful in Latin America. Local scholars and technocrats discussed the Gujarat reconstruction model, the policy, and the structures put in place by the government to deal with risks. Thiruppugazh unpacked how the disaster management policy in Gujarat was based on a clear identification of the role of government in protecting human beings, providing services, and guaranteeing security.

Not once during the conference in Gandhinagar was the term "resilience" mentioned, nor was it used in policy documents or strategies at that time. This was policymaking before resilience rhetoric. Officials instead emphasized the importance of strong public institutions, their responsibility for fair resource distribution, and protecting vulnerable people by avoiding their relocation to distant areas.

Citizens were actively involved in the postearthquake reconstruction of Gujarat. Responsibilities were shared, infrastructure was rebuilt, and strong institutions were formed. The authority implemented one of the most successful reconstruction programs in the world, and yet I never heard Thiruppugazh mention the word "resilience." He and his team had achieved remarkable results on their own terms (literally).

But things in Gujarat changed quickly. Thiruppugazh was promoted to higher office, and in 2014, Modi became prime minister of India. When new disasters occurred, the government could not replicate the success of the 2001 Gujarat State Disaster Management Authority.

A few years following Gujarat's reconstruction program, India joined other countries of the Global South in embracing the concept of resilience. When the Rockefeller Foundation offered US$160 million to institutionalize resilience worldwide, Gujarat stepped in. In 2007, the city of Surat in Gujarat was inducted as one of the one hundred resilient cities of the Rockefeller Foundation. That same year, UNDP issued a policy paper called "Towards a Disaster Resilient Community in Gujarat,"[7] which eventually led up to a conference on resilience co-organized by the government of Gujarat in 2008. The resilience jargon was widely adopted in policy documents by then. Even journalists started to explain how cities and territories needed to become more resilient.[8] Academic events on resilience took place,[9] and in 2015, a policy document sought "to enhance the State's resilience to cope with future earthquakes."[10] Modi's tweet was just the icing on the cake.

While the 2001 reconstruction in Gujarat demonstrated that India was able to accomplish great things without following the resilience guidelines and frameworks promoted in the West, the vocabulary had caught on. So, how did this obsession with resilience capture people's attention worldwide?

THE NEW FALSE IDEA

The original purpose of resilience was to recognize the power of ordinary citizens to overcome risk and destruction. It was initially a powerful idea meant to combat the victimization that could result from observing disasters under the lens of the vulnerability theory alone (see chapter 1). This compels us to shift our attention from people's weaknesses and fragilities to their strengths and capabilities. In doing so, we reveal their full potential rather than treating them as passive victims of an oppressive system. The initial hope for resilience was that government and charities would stop seeing citizens and beneficiaries as inactive receptors of aid. By highlighting citizens' agency, institutions would thereby recognize their power to change the current state of things.[11]

While "sustainability" became the most popular buzzword for reducing impacts *on* the environment, "resilience" emerged as a way to examine the impacts triggered *by* it. The metaphor was captivating. It insinuated that individuals and various social groups could "rebound" after a major tragedy. Much like plants and animals, it exhibited our capacity to adapt to hostile conditions and tackle adverse situations. The imagery of ecosystems was almost romantic. Instead of "breaking," or admitting defeat after a major event, there was hope that individuals and social groups could "bend" and recover in the wake of hazards. They could "go with the flow" during a catastrophe and recuperate afterward, *building* their way *back* even *better* than before.

Examples of resilience were also inspiring. Indigenous communities from different corners of the world, for instance, have been able to live in hostile environments for centuries. They have adapted and prevailed over hardships, often learning to live with them. Perhaps contemporary societies still have that capacity to confront disasters. It's just a matter of adjusting our approach in extreme situations—we can bounce back if we assert ourselves.

The notion of resilience became useful to explain why and how some individuals and groups avoid disasters. Indigenous communities and most preindustrial societies learn directly from their surroundings, coexisting with the forces of nature in a symbiotic relationship rather than fighting them. Vernacular solutions to housing, healthcare, food, and water provision show how we can grow stronger by "connecting" with our environment. Perhaps modern urban systems and industrialized societies can also stop rivaling nature and instead live in reconciliation with it. By focusing on people's "strong spirit of resilience," to paraphrase Modi, we could encourage and empower citizens to overcome tragedy and produce positive change. People and systems could recover faster and stronger. A fascinating concept, right?

Fascinating but flawed. As soon as resilience was propelled to stardom in academia and policy, it started to tumble.

The first flaw identified by scholars and practitioners was the idea of recovering the previous state—the predisaster conditions. The notion of "bouncing back" after a disaster may have been inspiring, but it was often contradictory. Why would a society want to go back to the situation that eventually led them to disaster? In many cases, recovering predisaster conditions didn't make sense. Given the chance, most Haitians, for instance, would not like to go back to the conditions of poverty and marginalization that existed *before* the 2010 earthquake. Similarly, black Americans would not like to reproduce the segregation, racism, and neglect that underscored the tragedy of Hurricane Katrina. As Kelman, Gaillard, Lewis, and Mercer concluded, this " 'return to normal' or 'back to normal' should perhaps not be part of addressing vulnerability and resilience."[12]

In observing this inherent contradiction, some intellectuals and consultants tried to rectify the message. They argued that the real objective of resilience was not to bounce back but to bounce forward—what sounds like a game of wordplay. The bounce forward principle implies that social groups must *improve* predisaster conditions during the reconstruction period. The real objective of reconstruction, therefore, hinges on the ability to enhance conditions and stop disasters from happening again. The problem is that this theory does not specify the type of change required after disasters (see the dilemma "They Want to Build Something Modern Here" in chapter 2). In many ways, this pivot brought us back to square one.

What exactly must be changed to prevent another disaster? Again, the answer was elusive.

The second flaw concerns the very core of system adaptation. The resilience metaphor is predicated on the conviction that human-made systems, such as neighborhoods, cities, infrastructure, and communication networks, behave like natural systems (*eco*systems). This suggests that human and social networks are programmed to restructure themselves after a major disruption, such as a disaster. They are expected to recover in the way that trees heal after a branch is cut and keeps its functions even after many leaves are amputated. But scholars soon remarked the limitations of this comparison. Hard infrastructure (such as a network of highways) is not "preprogrammed" to work after a major flood. It requires human beings setting up deliberate measures to recover it. Unlike *eco*systems, several systems created by humans still depend on individuals making premeditated decisions and taking effective actions. Their capacity to cope with adversity is still framed by the effectiveness and agency of individuals who are vulnerable.

Surely, parallels can be drawn between ecosystems and the outstanding autonomy, decentralization, complexity, and flexibility of modern communication systems, such as the internet and cell phone networks. But most ports, train corridors, airports, houses, sewage and other man-made systems still lack the same level of adaptability that characterizes the web or telecoms. "That social systems are akin to ecological ones may have been the idea that gave birth to human ecology and cultural ecology," explains David Alexander, a professor specialized in disasters at University College London. But he argues, "this does not mean that social interaction is fundamentally the same as the ecological pyramid of species and trophic levels."[13]

Techies fought back, asserting that artificial intelligence would be able to replicate nature's autonomy for recovery. An intelligent building or neighborhood would not need a human being to reprogram activities and resume functionality, especially since algorithms and electronic systems can provide this level of self-recovery. And yet, most individuals who followed resilience had been inspired by indigenous communities and vernacular solutions. They had simple ideas in mind—not robots. Besides, electronic systems require energy. Wasn't it contradictory to achieve resilience through electricity and megabytes, forms of power that frequently fail? This high-tech brand of resilience has become puzzling for many.

The third flaw concerns the value attributed to persistence over time and after adversity. Common interpretations of resilience position this ability to rebound and endure as a definite advantage. Continuity is unquestionably desirable. But experts soon noted that this standard is absurd.[14] Several wrong practices such as violence, exploitation, and corruption have the capacity to persist, and there is nothing good in their resilience.

A permanent value attached to strengthening the status quo is also beneficial to political and economic elites. In most cases, the most powerful adopted resilience as a way to maintain "business as usual" or to justify their privileges, ensuring that they remain untouched. This moved many critics to wonder: Resilience of what, to what?[15]

But again, answers to these questions proved to be unattainable, because they require tangible measures. For example, how could one predict the extent of an infrastructure's resilience in the face of a volcano eruption? How can we know if a company or city is becoming increasingly resilient to an unknown pandemic or unanticipated terrorist attack? How do we determine whether an investment in resilience is producing the expected results? For two decades, scholars and costly urban consultants have been trying to quantify resilience. They have created dozens of frameworks and checklists to assess adaptation and create benchmarks for policy monitoring, though few of them are actually put in practice other than for selling more urban consulting services. Tracking change in a meaningful manner is expensive, which results in oversimplifications, abusive shortcuts, and obvious generalizations. Despite many efforts, resilience is still significantly difficult to measure.

After perceiving common failures in climate change and disaster adaptation, scholars have adopted increasingly abstract terms. "Sustainable adaptation" is now used to distinguish the types of change that scholars strive for from the results eventually obtained. Other terms distinguish "specific" and "general" forms of resilience.[16] And this vocabulary only continues to expand within academic circles: "adaptive capacities," "holistic resilience," "climate-proof solutions," "elastic adaptation," "adaptive cycles," "basins of attraction," "panarchy," "bounce-back answers," "increased flexibility," and "critical adaptation" have become common terms found in the "guidelines," "roadmaps," and "pathways" of international agencies aiming to institutionalize resilience. Terms multiply, but their meanings remain just as illusive. As summarized by James Lewis, a researcher from Datum International, "Current interpretations of social resilience do not match reality."[17]

Resilience has often been considered a crucial component of sustainability. According to Karen I. Sudmeier-Rieux from the University of Lausanne, resilience still "needs to be assessed critically as one attribute of sustainable development, not as a lesser substitute."[18] As a matter of fact, most authors have failed to recognize the innate contradictions between resilience and sustainability. To fully appreciate this discrepancy, we must examine how infrastructure impacts our daily lives.

THE MOST RESILIENT HOUSE IN NORTH AMERICA

In January 1998, a series of ice storms hit Eastern Canada and the United States. As much as 100 millimeters of freezing rain covered streets, homes, electric towers, wires, train tracks, and virtually all other infrastructure. More than 4.5 million people lost electricity in the middle of winter, when heating is of prime importance. Hundreds of thousands of Canadians were in the dark for weeks. It is estimated that thirty-five people died, many of who were intoxicated by carbon monoxide as they tried to prevent freezing by lighting fires. Many others died of hypothermia, and over six hundred thousand people, mostly seniors, were relocated to hotels, hospitals, and temporary shelters. Some three hundred thousand farm animals also froze to death, as farmers failed to keep their barns heated.[19] For several days, the majority of electrical systems collapsed in the large Canadian province of Quebec.

In 2003, about fifty million people lost power for two days straight in what was the biggest blackout in North American history. The event led to eleven deaths and resulted in losses estimated at US$6 billion.[20] The New York City subway, communications, and several airports and trains were inevitably interrupted.[21] The city was in complete darkness for several hours.

These two events are recent reminders of our dependence on the infrastructure that we often take for granted. Unsurprisingly, for many people in the developed world, living *off* the grid is the ultimate resilience experience. But being disconnected from public services is still a tragedy for millions of people in poor countries. Despite that, those who are truly disconnected from public services in poorer countries struggle tremendously, and people in rich nations still dream of moving to remote locations and experiment with autonomous homes. These homes typically have local water and sewage arrangements and are powered by solar panels and geothermal systems (a relatively autonomous heating and cooling mechanism that transfers heat to or from the ground). This tendency has also inspired millions to adopt small-scale agriculture, aiming to become less dependent on commercial food chains. Today, there are hundreds of youtubers, bloggers, and writers that proclaim their off-the-grid gardening experiences as resilient practices to be adopted at large.

One of these "resilience explorers" is Albert Hamilton.[22] After suffering the Saguenay floods in 1996, the 1998 ice storm in Eastern Canada, followed by the loss of his own home that burned down in 2013,[23] Hamilton finally decided to quit his practice as a building contractor and build a resilient home for his own family. Hamilton's new home seems to have a solution to any possible difficulty. No running water? The house is furnished with a well that recuperates rainwater. Lost electricity? Hamilton's home consumes minimal energy, with photovoltaic panels to heat the house and water. Lack of food in a major crisis? Well, the home has a hydroponic garden in the living room, where Hamilton is able to grow food. Flooding in the area? This resilient home is built on high ground to prevent disaster.

"The house is connected to the grid," admits Hamilton, but it has sufficient "redundancy in systems, which makes it almost autonomous." The house is almost 2,000 sqare feet and costs about US$500,000. Not quite a small or affordable home, but Hamilton's home has been hailed as "the most resilient house in North America" by the founder of Building Green Inc. and the director of the Resilient Design Institute.

What would happen if more people followed Hamilton's example? Would our society become more resilient?

Quite the opposite is true. In fact, producing more homes like Hamilton's would lead to environmental catastrophe.

In order to be resilient, Hamilton's home relies on redundant heating and cooling systems, overdimensioned walls, triple-glazed windows, and other sophisticated components. There are, for instance, batteries to stock energy, a wood stove, and a back-up generator. The home needs ample sealing material to keep it almost totally hermetic—not to mention all the energy required to manufacture, transport, and assemble so many of these components. Hamilton might be paying a lower electricity bill than his conventional neighbors, but he has spent more energy to build his home than the average person. In fact, Hamilton's home in Canada is more polluting than it first appears. The electric grid that powers his neighbors is produced by dams, which means it generates very little carbon emissions, whereas his photovoltaic panels are quite polluting. If more people followed Hamilton's example, vast amounts of land would be needed to install their solar panels. The footprint of each home would be much higher than any conventional unit.

By farming in his resilient home, Hamilton attempts to become a locavore—but what would happen if we all took his approach and opted for only locally-produced food? Well, some experts believe that this would also be disastrous. The amount of energy required—and carbon emissions generated—to produce Hamilton's tomatoes in his home garden are enormous. His hydroponic tomatoes might be tastier and have fewer traces of pesticide than the average. In harvesting them, Hamilton might *feel* like he exudes the essence of survival. But these tomatoes make little environmental sense at a larger scale. Simply put, the food produced by Hamilton requires more energy, and thus releases more carbon emissions, than the vegetables that his neighbors buy at the local supermarket.

This is a common trend assessed by James E. McWilliams, a professor at Texas State University. In *Just Food: Where Locavores Get it Wrong and How We Can Truly Eat Responsibly*, McWilliams dispels myths about locally produced food. For instance, McWilliams explains that reducing "food miles," or food transportation, "makes little progress towards the ultimate goal of sustainable production."[24] David Owen, the journalist I introduced in chapter 3, states, "How far you live from your grocery store is of far greater environmental significance than how far you live from the places where your food is grown."[25] So if we decided to move to a remote location and transform millions of hectares of agricultural or wild land to build our autonomous homes, our carbon footprint would explode. Hamilton's home has a rural, lake-side location, with two cars, a snowmobile, and a boat.[26] Even with fewer vehicles, his needs for transportation would still produce considerable carbon emissions.

For buildings to become resilient, they depend on back-up systems and extra provisions. These resilient systems need energy for production, transportation, and assembly that is not required in conventional solutions.

While Hamilton's home is experimental in his effort to do the right thing, his actions have very little environmental value overall. His approach to resilience is at odds with his sustainable intention.

Many people argue that these limitations are only temporary, and that once technology improves, resilient solutions will prevail. This might be true, but it also points to the enormous tasks we have ahead. To overcome the flaws in Hamilton's home, we would need to produce, distribute, and stock vast amounts of clean energy, which is still far out of reach.

If people like Hamilton truly wanted to reduce their impact on the environment and avoid disasters, they would make a different decision. They would perhaps move to an apartment half the size of Hamilton's home in a midrise building in a dense city. This might not sound very green, and yet, this would immediately reduce transportation, all while optimizing existing infrastructure. A smaller unit would save energy for lighting, cooling, and heating. The smaller apartment size would perhaps prompt Hamilton to accumulate less and incite him to buy less furniture, decoration, appliances, vehicles, and other goods. It would also mean that he would use fewer pesticides and less water to maintain the lush lawn of suburban and rural houses. Moving further from the wild is still our most resilient action.

This contradiction between the objectives of sustainability and resilience is both common and counterintuitive. Resilience arose in public policy as the offspring of sustainable development. Hamilton —like most politicians and professionals—capitalized on resilience as a way of pursuing a green agenda that avoided permanent destruction. And yet, as Hamilton's home demonstrates, there are significant discrepancies between the objectives of resilience and those of sustainability.

A paper published in the prestigious journal, *Environment and Planning*, concludes that resilience "leaves wider sociocultural concerns unaddressed." Instead, it "emerges as a narrow, regressive, techno-rational frame centred on reactive measures at the building scale."[27] This is the issue that underlies most people's approach to resilience. We have disconnected the technical aspects of human systems from their sociocultural context, thus failing to detect the gap that exists between sustainability and resilience.

Another problem is that whereas most Ted-talkers, successful politicians, and celebrated scholars are now fluent in resilience rhetoric, not everyone on the ground speaks the language.

WE SAID, THEY SAID

In 2017, my research team obtained a grant to investigate how low-income people in Latin America and the Caribbean were coping with global warming. We selected five cities in Cuba, Chile, Haiti, and Colombia for our study, focusing on low-income neighborhoods. There, we conducted a series of interviews and focus groups with local residents. We asked them about their own resilient capacities and how they have adapted to the risks of an increasingly warming planet.

Our team quickly realized that none of the individuals we spoke to in Cuba or Colombia used the terms we had incorporated into our project brief. Community leaders in focus groups did not refer to "resilience," while disaster-affected residents almost never discussed the notions of "vulnerability" or "adaptation." When ordinary citizens used the Spanish equivalents of those terms, it was only in strained responses to questions or comments in which they were already embedded. We then suspected that respondents adopted this lexicon and mirrored our own sentences to better understand what we meant by them.[28]

For instance, when one of our interviewees wanted to describe her neighbor, Mariana, she would not say, "Mariana is resilient." She would instead say something like, "Mariana has the right attitude to overcome any challenge." We originally interpreted this as a form of resilience in our research notes, but after documenting longer narratives, we recognized that such loose translations distorted the overall message. For example, we were assuming that certain activities or people were "resilient" simply because we had already subconsciously viewed them that way; not because residents were implying it.

Surely, low-income Cubans and Colombians would like to improve their living conditions. However, we noticed that by asking them questions deriving from academic jargon, we were leading them to adopt predetermined ideas. We wondered: Were we studying a subject that no one on the ground was concerned about? Were we patronizing for imposing a new language on local residents—or even worse, enacting a new form of intellectual colonialism imposed by the North on the Global South? Had we fallen into the role of urban consultants such as those in the Rockefeller's 100 Resilient Cities program?

Perhaps the concepts that intellectuals and ordinary citizens engage with are fundamentally similar. In this case, my team and I could match and translate terms, applying our knowledge to bridge vernacular ideas and theoretical concepts. Another possible situation is that the terms adopted in academic and political jargon are appropriate, and the ones used by ordinary citizens are imperfect or rudimentary versions of them. If this is true, intellectuals could elevate raw ideas to a higher status of abstraction, conceptualization, and generalization.

These two scenarios proved to be false. They implied a form of parallelism between vernacular and technical narratives. In the first case, it is a matter of finding equivalents, much like a translator vacillating between languages. In the second case, it is a matter of formalizing concepts so that rough material can be used systematically in science.

We eventually found that a third situation was more plausible. This kind of parallelism didn't exist. As a matter of fact, the majority of citizens were rather confused by the notion of resilience. They were more occupied by the pragmatic challenges they dealt with on a daily basis rather than their adaptive capacities.

We initially thought that citizens were going to express their concerns with floods, longer periods of rain, and harsh periods of drought. But more often, they wanted to talk about their unpaved roads, the difficulties of finding a job, and their continuous struggle to make ends meet. They were certainly anxious about natural hazards and environmental degradation, but these challenges were typically framed within an explanation of daily struggles that, at first, seemed mundane to us. When we began to discuss the recent increases in precipitation, for example, residents quickly reminded us how difficult it was to buy groceries in their neighborhood. When we spoke about the alarming frequency of flooding, they addressed their lack of health care insurance.

We concluded that we had fallen into the resilience trap. Much like technocrats, charities, and government officials, we had absorbed the resilience jargon without noticing how little it resonated with regular citizens. When we explained this to our other colleagues, they told us that they found similar patterns while working in low-income neighborhoods across Europe, Asia, and the United States. We learned something relevant about global warming: climate change and the challenges it poses are part of the regular struggles that citizens face. The new risks that emerge in a warming planet cannot be understood separately from the daily toil of poverty, marginalization, racism, and neglect. Environmental challenges and social inequalities are intrinsically linked. They are simply two faces of an ongoing fight for social and environmental justice.

We saw in chapter 1 that, for several decades, Cuba had developed a comprehensive risk-reduction plan. This plan made Cubans less prone to the effects of disasters.[29] It hinged on the government's responsibility for protection, safety, public health, education, and information. It was also based on the authorities' engagement in immediate recovery, and later, reconstruction efforts. For almost forty years, Cuban risk-reduction and prevention has been portrayed as a duty of state protection rather than a "resilient approach." Similarly, disaster risk-reduction plans adopted in Colombia before 2010 almost never used the term "resilience," but as in Modi's India, pressure is mounting in Latin America and the Caribbean to finally adopt resilience.

The Colombian cities of Cali and Medellín are now part of the Rockefeller Foundation's 100 Resilient Cities. In 2015, the United Nations Office for Disaster Risk Reduction and the local organization Corporiesgos signed an agreement to transform the Colombian region of El Valle into a resilient territory. UNDP officially declared Colombia a "resilient territory" in 2018, and now local newspapers in Cali, such as *El País*, regularly publish articles about resilience, making extraordinary efforts to explain the term in plain Spanish.[30]

The same pattern emerges today in Cuba, where resilience has become part of the jargon of politicians and technocrats. UN Habitat, UNDP, and other international consultants are largely responsible for this shift in narrative. In 2012, UNDP partnered with local institutions to create a plan founded on "resilient cities." It also published the guidelines for the reconstruction of "resilient housing" in the city of Santiago. Three years later, UN Habitat and UNDP partnered with the National Planning Institute (Instituto de Planificación Física) and the Civil Defense to conduct a project on "urban resilience in Cuba."[31] This led to the publication of the *Guide for Urban Resilience*, a document that stipulates how and why municipalities should adopt the resilience framework.[32] Conferences and seminars followed, including a 2018 debate on urban resilience organized by UN Habitat.[33] Resilience jargon has since been applied in energy, agriculture, and forestry sectors. In 2019, UNDP financed a project to install photovoltaic panels in Cuban homes, an attempt to "increase resilience among communities in the face of extreme meteorological events, through the use of energy renewals."[34] The following year, it launched a climate change program called Coastal Resilience.[35] Today, there are dozens of projects funded by United Nations' agencies dedicated to fortifying resilience on the island.

This change of narrative in Latin America is deeply disconcerting for two reasons: First, by embracing resilience, environmental agendas become disconnected from people's needs and their calls for social security, stable jobs, equality, and anticolonialism. The climate agenda has often looked ahead at the challenges that societies will face in fifty and one hundred years, but these plans have no chance of succeeding if they continue to ignore people's basic rights and daily needs. By focusing on mid- and long-term goals, they overlook today's most critical battles.

Second, a climate change agenda that centers on human adaptation to meteorological hazards fails to prioritize social justice—especially in developing countries. "The current program of 'mainstreaming' adaptation into existing development logics and structures perpetuates an antipolitics machine," states Scoville-Simonds, Jamali, and Huffy—three climate change scientists. This obsession with adaptation is "obscuring and depoliticizing rather than addressing the political dimensions of the adaptation problem."[36] Kelman and Weichselgartner conclude that "the apolitical ecological resilience thinking tends to favour established social processes and traditional societal structures at the expense of social transformation."[37] Similarly, Silja Klepp and Libertad Chavez-Rodriguez, two researchers from Kiel University, Germany, and the Centre for Research and Advanced

Studies in Social Anthropology in Monterrey, Mexico, affirm that "despite its significant political effects, most of the discussions concerning 'adaptation' are effectively framed in an apolitical manner."[38]

Unfortunately, politicians' climate agendas, in addition to the consulting services offered by international agencies, typically avoid any mention of enlarging the welfare net. Instead, international urban consultants and cities too often exploit resilience—and the associated concept of adaptation—to legitimize transfer of responsibilities to people and businesses.[39] In this way, resilience has enabled urban consultants to charge high fees over the past twenty years, while doing very little to reduce the vulnerabilities of the poor and marginalized. "The current emphasis on adaptation as a trope implies accepting a world in which disturbance and crisis are constant features," argues Greg Bankoff, a professor at the University of Hull. The notion of adaptation leads to "a continual need for neoliberally sanctioned discourses about resilience and change. It positions disaster as an endemic condition [so that] society must remain in a permanent state of high alert." As Bankoff continues, this "obscures questions about the role of power and culture in society, and about whose environments and livelihoods are to be protected and why."[40]

YOU MUST BE RESILIENT, NOW!

In 2007, floods affected about fifty-five thousand properties in the UK. Inundations repeated in 2012 and 2014, and by winter 2015, floods and other damages associated with storm Desmond in the UK were estimated at US$1.6 billion.[41] Newspapers then reported that about ten thousand homes were being built on floodplains each year in Britain.[42] This was no exaggeration: recent studies found that as much as 27 percent of new housing in England is built in flood hazard areas.[43] In 2015, another study found that 90 percent of floodplains are no longer able to withhold water in Northern England.[44] This means that water now flows downstream more quickly, causing disasters in suburban and urban areas.

Most flood-prone residential areas in the UK are protected with water defenses and levies. But urbanization and climate change are putting pressure on towns and villages, where "flood defenses are unable to cope."[45] A study conducted by the UK Committee on Climate Change concluded that, between 2001 and 2011, the rate of urban development in floodplains was higher than in safer (dryer) locations. Within that period, more than 40,000 properties were built in areas with a high level of flood risk, and about 70,000 in areas with a moderate risk.[46]

Why do authorities tolerate so much construction on floodplains? Researchers point to various explanations, including poor urban planning, insufficient enforcement of restrictions, lack of information about flood risks, dubious arrangements between residential developers and municipalities, and overconfidence in water defenses. Others have noted that, because such a significant portion of land in the UK is flood prone, a certain level of risk must be accepted.[47]

But in 2014, I and my colleagues from Loughbourough University—Ksenia Chmutina, Lee Bosher, and Andrew Dainty—implemented a study to further investigate those claims.[48] In addition to speaking with a dozen officers responsible for disaster management, our team analyzed about thirty policy documents devoted to disaster risk prevention and reaction. We found that, at the core of risk tolerance in the UK, lies a standard for how resilient citizens are expected to be.

As a matter of fact, government policy surrounding flood risk in the UK is based on the premise that residents are (or must become) highly resilient. According to the UK government, households and businesses are expected to take "the appropriate steps to better protect their properties through property-level resistance and resilient measures."[49]

Here, resilience is less of an explanation for how people *cope* with adversity and more of an expectation for how they *should act* according to authorities. "There will be times when individuals and communities are affected by emergencies that are not an immediate danger," explains the UK Strategic National Framework on Community Resilience. These individuals and social groups "will have to look after themselves and each other for a period until the necessary external assistance can be provided." The policy goes on to specify how people should act: "Communities will need to work together, and with service providers, determine how they recover from an emergency."

However, this does not mean that citizens are free to make their own decisions. Instead, they are delegated activities and responsibilities based on what the authorities have decided is best. As one researcher summarizes, the UK government puts "local people in the driver's seat, when in reality, the direction of their journey has already been decided." People resent the fact that they are delegated activities and expenses but are refused decision-making power.

Given this scenario, there is now a widespread perception that the resilience agenda in the UK operates to fortify neoliberal principles. The idea here is to move away from state-enforced security, transferring responsibilities to individuals and businesses. This policy is implemented by delegating responsibilities from the national government to regional and local ones, and finally, to individuals and enterprises.

Our study at Loughbourough University eventually concluded that "resilience is a politically charged term used as a tool in the government's attempts to keep centralized control," while simultaneously "shifting responsibilities from the State towards local authorities, communities, and businesses."

Unfortunately, the resilience scheme in the UK is not the only one. Other scholars have noted that politicians, charities, and companies around the world have found a way to harness the principle of resilience to advance their own agendas.[50] Resilience has become an ideal tool for manipulation. As Michael Watts explains, it forms "governable subjects" and "provides a powerful anticipatory calculus, one of a flotilla of technologies associated with a security assemblage."[51]

Once hijacked by authorities in the UK and many other countries, resilience is used to legitimize government disengagement with the most vulnerable while serving the interest of economic and political elites. Evans and Reid observe that "the representation of Indigenous peoples as possessing exceptional capacities to care for their natural environments, to adapt to climate change, and deal with extreme weather events has become a governing cliché of Western neoliberal governance." The danger is that the real purpose of this approach is "to discipline Indigenous peoples themselves into performing their own resilience."[52] Social resilience, explains Greg Bankoff, "has become a core constituent of the neoliberal economic agenda now expressed in terms of sustainable development and its prescriptions for institutional reform."[53] As such, Michael Watts identifies "adaptation, security, risk management, and resiliency" as "contemporary hegemonic forms in which particular forms of life constitute the basis of neoliberal rule and governance."[54]

We have widely adopted the language of risk, yet have failed to detect the risk of adopting this language. "Labeling people as 'resilient,' " remark Ilan Kelman, Jessica Mercer, and J. C. Gallard, "might also mean that they believe it." By adopting resilience, "we need not worry about changing anything. . . . We are so resilient that we should take care of ourselves."[55] Seeing as there are so many cases where resilience is manipulated for political gain, academics have come up with a new term for it: maladaptation.

RESILIENT, BUT ON MY OWN TERMS

Although resilience has become the primary language of politicians, we know that researchers are increasingly challenging its substance. One of them is Isabelle Anguelovski, a Spanish researcher at the Universitat Autònoma de Barcelona, affiliated with the city's Lab for Urban Environmental Justice and Sustainability. A few years ago, Anguelovski explored policy changes introduced after Hurricane Katrina in the United States. She found that they were facilitating "post-Katrina gentrification" of neighborhoods along the Mississippi River's natural levees. They also largely failed to address "persistent and uneven flood vulnerability through land use reform." The consequence of this is that "poor African American communities will likely continue to be disproportionately vulnerable."[56]

While investigating the case of Surat, she identified a similar trend. Following the Rockefeller Foundation 100 Resilient Cities endorsement, several policies were implemented there—some of which Anguelovski and her colleagues found "catered to the interests of vulnerable communities." But these policies were also "entrenched by ongoing privatization trends" that redirected infrastructure toward economically important industries, while monetizing water and sanitation services and evicting marginalized communities. Due to relocation and new housing constructions, conducted in the name of resilience, units were often located far from jobs.

You might suspect that these negative effects could have been avoided if interventions focused on enlarging green areas instead. What can go wrong with additional trees, new patches of nature, and green zones?

The answer is a lot, according to Anguelovski. She sees how green infrastructure has produced secondary effects and resistance in cities such as Metro Manila and Medellín. Surely, Metro Manila's drainage canals and Medellín's new green belt (see chapter 7) were presented as win-win adaptation actions. But Anguelovski is aware that they also led to displacement of low-income communities. In many cases, poor families were immediately evicted to free areas for environmental protection. She also fears the effects of what she calls "climate gentrification," or the displacement that comes from rapid increases in rents and status, ironically caused by attempts at reducing risks in the face of global warming.

Another scholar challenging common narratives of resilience is Lizzie Yarina, an American researcher with the MIT Urban Risk Lab and a Fullbright fellow in New Zealand. In 2018, Yarina conducted a study on climate change in Asia. She explored how four large Asian cities developed risk-reduction plans under the principle of resilience. In Metro Manila, her findings confirmed those of Anguelovski, and in Jakarta, she investigated an initiative that aimed to make "room for the river," called the Jakarta Urgent Flood Mitigation Project, supported by the World Bank. This project relies on removing 2,300 housing units and evicting some two hundred thousand people, mirroring patterns of eviction for a project planned for Ho Chi Minh, Vietnam. According to her study, about 1,500 people will be evicted because of this initiative. More than this, the project is likely to increase flood risk for the poorer, rural neighborhoods (see a similar case in chapter 5). Another river beautification project that requires massive displacement was planned for Bangkok, Thailand. Yarina concludes that "the rhetoric of climate adaptation is doublespeak for the displacement of poor, informal communities, and an alibi for unsustainable growth." For her, maladaptation is "a situation in which projects undertaken in the name of climate change adaptation are overrun by interests other than the stated or intended objectives of the climate-change adaptation project."[57]

I do not have evidence of how residents reacted to these specific initiatives reported by Anguelovski and Yarina. Many might indeed have benefitted from the changes that were made. But results from my own research in Latin America prove that residents often resent infrastructure and urban transformation projects that, through relocation or rapid changes in housing conditions, alter their social networks, disrupt their access to jobs and services, and fail to address their main concerns about crime, violence, unemployment, and access to health and education. This resentment emerges even when initiatives are made in the name of resilience.

A CLIMATE (AND CONSULTING) AGENDA DISCONNECTED FROM PEOPLE'S NEEDS AND EXPECTATIONS

We have now seen how the climate agenda has become increasingly disconnected from the needs and expectations of the have-nots and most vulnerable. In many cities, this agenda has been structured around the interests of the political and economic elites. This is particularly worrisome, given that the poor and most vulnerable are most likely to be impacted by the effects of global warming.

Quite often, the climate agenda promoted by urban consulting agencies has introduced global warming as an environmental and atmospheric emergency, disconnected from the social struggles and requirements of vulnerable social groups. A 2019 national survey conducted by Yale University and George Mason University confirmed this disconnect in the United States.[58] While 75 percent of Americans see global warming as an environmental issue, and 64 percent as a severe weather problem, only 29 percent see it as a poverty issue, and 24 percent see it as an issue of social justice and fairness. Ultimately, 54 percent of Americans view global warming as an economic issue and only 38 percent as a moral one. Decision makers and academics have reinforced this separation between the environmental crises and social and moral affairs, highlighting the profound consequences of the climate agenda. "The Climate Adaptation Strategy" argues Yarina, "is filled with ideas that may make sense in a Western context but fail to translate smoothly to local conditions."

Is this criticism exaggerated? Many defenders of resilience believe it is. Some scholars and practitioners recognize that the resilience agenda is an imperfect tool but claim that it is also a harmless one. They recognize that it sometimes fails to achieve the expected results, yet they contend that its attempts to create positive change are better than nothing. This positions resilience as a value-neutral idea that might produce positive effects but is unlikely to produce negative ones. Environmentalists argue, for instance, that while resilience might not produce the ambitious changes originally anticipated, it can at least generate environmental and risk awareness. For years, the Rockefeller Foundation's 100 Resilient Cities program was relatively inefficient, yet it created awareness among municipal employees and technocrats about environmental challenges and destructive events.

The cases we have seen in this book, however, demonstrate that resilience is hardly a value-neutral idea. In fact, by introducing the resilience agenda in cities and social groups, consultants might introduce problems that did not previously exist locally. They might also distort the messages and claims that are typically voiced by historically marginalized citizens and social groups. A climate agenda based on the premises identified by urban consultants in New York, Washington, London, Barcelona, and Nairobi, are too often disconnected from problems on the ground and thus overlook expectations and aspirations that are important to the citizens at hand. If we ignore the relevance of their daily struggles and challenges, the climate agenda then produces solutions that do not actually address the needs and expectations of the most vulnerable. If we don't do a better job at integrating social and climate agendas, writes Klepp and Chavez-Rodriguez, actions will remain "concentrated in the local, as if adaptation capacities were only a matter of enhancing local resilience, targeted through technical intervention and expertise by Western consultants."[59]

By adopting the resilience discourse without questioning its value, professionals in engineering, architecture, urban planning, design, and landscape have perpetuated a body of knowledge that fails to capture the realities of the most vulnerable. The same applies to academics and researchers, who have adopted the resilience ideology without challenging its value. This has increased the gap between academic work and the urgent problems we face today.

How difficult is it to override the resilience framework? Very hard, you may conclude if you decide to read Katrina Brown's book *Resilience,*

Development and Global Change. Over the course of about 184 pages (in my Kindle version), Brown, a professor at University of Exeter, provides a comprehensive criticism of resilience. Page after page, you learn how resilience is too abstract, too disconnected from real-life experiences, too manipulated, and too abused. While discussing the use of resilient frameworks and conceptualizations in development policy, Brown argues that "they adopt a new lexicon and some new concepts, but still do not address fundamental causes—of poverty, inequality, global change or vulnerability."[60] You can only agree with her and hope that she provides a radical way out, but in the last twenty pages of Brown's book, she reveals that the very framework she has carefully criticized can still be useful in development policy. Brown refuses to abandon the notion of resilience and instead revisits it from a political ecology perspective. She argues that resilience must be based on three pillars: rootedness as an attachment to place and careful consideration of meanings and local action; resistance as sociopolitical reaction to power systems; and resourcefulness as the capacity to mobilize resources and skills.

Brown's intention to dismantle resilience does not go far. And this is where I wonder: Why do we need resilience jargon to talk about rootedness, sociopolitical resistance, resourcefulness, the protection of nature, and social justice?

In order to succeed, risk reduction and the climate agenda must be framed within the broader strife for social and environmental justice. New concepts and frameworks are perhaps needed to respond to the urgency of global warming, but they must be built from the ground up.

THE MORAL VALUE OF RESILIENCE

These challenges bring us back to the book I decided not to write. The strength, courage, and capacities of citizens must be recognized and celebrated. This is the first step to create effective climate and disaster-reduction policy. Human capabilities to solve major problems, cope with hardship, and overcome major disruptions must also be admired and documented. Thanks to these strengths, most cities are safer places today. Public health and natural hazards that caused millions of deaths and suffering years ago are less dangerous, and fewer people now die from diseases, fires, car accidents, floods, and poisoning (not counting drug overuse). We must study these successes, and what led up to them, to learn how to deal with current threats, especially those fuelled by global warming. This is a worthy subject for many books.

More importantly though is the need to act now. Preventing the rampant destruction of nature in contemporary societies is urgent, and I believe that the urgency of response trumps complacency. People's capacities, along with social progress, should not be used to justify disengagement with the needs of our most vulnerable. The recognition of individual agency does not excuse approaches where government responsibility becomes diluted. A risk-reduction agenda cannot derive from our human potential to overcome tragedy. It must stem from engaged dialogue about our moral responsibility to ensure disaster prevention and avoid disaster creation.

Resilience, I believe, fails these moral tests. And much like sustainability, it is too compromised to be saved. Instead of relying on the muddled and manipulated idea of resilience that urban consultants are selling us, we can resort to the basic sources of well-intentioned action, which requires a thorough evaluation of the moral values underlying our perceptions of risk. I will return to this subject in chapters 7 and 8, but for now, let's explore our optimism in community participation.

5 Participation

"They Want Us to Participate in the Construction of I-Don't-Know-What"

There were disturbances, there were regrettable instances of bloodshed, but the masses of Babylon, at last, over the opposition of the well-to-do, imposed their will; they saw their generous objectives fully achieved.

—Jorge Luis Borges, "The Lottery in Babylon"

WHAT THE MAYORS WANTED

"Money, so I can pay for a secretary," said one of the mayors. "Resources," intervened another, "to cover a technician's salary in city council." Then a few other mayors chimed in, asking for help to fund administrative support staff.

As the moderator of this workshop, I felt uncomfortable, to say the least. I suspected that these Haitian mayors were not going to get any cash at all, though that's what they truly needed. It took a couple of hours of discussion to reach this outcome—a puzzling one, as it proved to be.

Among the attendees were ten Haitian mayors, four deputy mayors, and four representatives of other municipalities. Only five months had passed since the Goudougoudou, the powerful earthquake that hit Haiti in January 2010 (see chapter 1). All of the cities represented at the event had been affected by the earthquake in some way. Many of them belonged to what is known as metropolitan Port-au-Prince, a conglomerate of municipalities with a total of four million inhabitants.

We were working from an economy hotel in the state of Louisiana, where the mayor of Lafayette hosted this workshop. The gathering was organized by the International Association of French-Speaking Mayors (IAFM), an organization that brings together the heads of 307 cities across fifty-two countries and regions—many of them former French colonies. The official objective of the two-day workshop was to draft a collective plan to assist Haitian mayors with postdisaster reconstruction. Participatory exercises had been planned for this purpose.

To be clear, the city of Lafayette and the IAFM were also extending their full support to Haiti's mayors during this harsh time. Lafayette had been greatly affected by Hurricane Katrina in 2005 and now arose an opportunity to transfer lessons learned. As such, I had been asked by the IAFM to organize an afternoon panel to help identify the mayors' most pressing needs. My first question for them—I thought—was relatively straightforward: What do you need to facilitate recovery?

The workshop's participants were not as naïve as I was. This was one of my first activities as a workshop animator. But many of the mayors present were highly experienced in participatory activities and were perhaps accustomed to false promises. It took a while to get the ideas flowing, but once the issue of financial resources for municipal administration came up, the conversation finally progressed. Each mayor described, in meticulous detail, the difficulties of trying to get things done in public administration, where computers had not been updated for years, facilities fell to ruin, and the few personnel members employed were underpaid and unskilled.

I reported the results of the panel to the IAFM officers. The key takeaway was that Haiti's mayors needed financial support to reinforce public administration. I quickly realized that this was not the answer the organizers wanted to hear, as they provided explanations to me that revealed another agenda. The IAFM wanted to organize an association of Haitian mayors so that their cities could assist each other during the reconstruction period. A new association, the officers explained, could distribute information and share new knowledge. It could also persuade elected officers and technocrats to exchange best practices for better coordination among cities. Strategic projects for the metropolitan area could be organized in joint collaboration, and besides, political pressure could be maximized to influence the national government. A sound idea, I thought.

The following day, when the IAFM representatives offered to create the association of Haitian mayors, the mayors' response was rather elusive. Some mayors returned to the idea of financial support for staff, while others went over the details of projects with insufficient resources, procedures, and administrative support. They seemed to have little interest in the new multicity association. That's when the organizers pushed a bit more to unpack the benefits of such an association. After hours of back and forth, one of the most senior mayors raised his hand—seemingly exasperated. He pointed out that this kind of association *already* existed. The problem was that this association was largely inoperative due to the municipalities' lack of staff and employees. The few workers employed were already overstretched across several projects and activities, so asking them to get involved in a multicity association was impossible. The room went silent.

When the existence of the Haitian association—and the causes for its lack of success—were revealed, the workshop lost its raison d'être. The Haitian mayors wanted financial resources, not another bureaucratic institution. The IAFM, meanwhile, had the capacity to support the creation of a new institution, not cash.

This workshop had cost tens of thousands of dollars. It had taken valuable time from leaders who were trying to recover their cities after the tragic earthquake. And yet, the workshop hadn't really achieved anything. It became clear to me that the event had not been organized to learn what the mayors needed—it was planned to rally their support for a new Haitian association. But nobody bothered to inquire whether such an association already existed. They assumed that it didn't and proceeded with their plans, which I had inadvertently played a part in.

The majority of the present mayors did not bother to attend the event's afternoon session. In the evening, I returned to Montreal knowing that I had failed to reach my objective and had essentially let down the Haitian leaders.

This story has two caveats. First, I believe the IAMF acted in good faith. Their intention was noble: to install a functional multicity association that would establish shared priorities and initiatives among mayors to reinforce the Haitian state. Second, corruption in Haitian municipalities is a major problem, so it is likely that Haitian mayors are involved in political games and pursue partisan interests. But even considering these two aspects, I couldn't help feeling used. I had agreed to organize this panel without knowing the underlying agenda. The mayors had similarly been invited without a clear sense of purpose in the organization's final aim. I don't know whether I could have made things better if I had known the real purpose of the workshop in advance. Perhaps not. Either way, I sure felt embarrassed.

THE KEY TO SUCCESS IN TIMES OF DISASTER

In 1982, the United Nations Disaster Relief Organization, known as UNDRO, published an influential report. UNDRO had noticed that for more than a decade, housing reconstruction projects led by governments and charities consistently failed to respond to the needs of affected communities. Houses and infrastructure were abandoned. Budgets and schedules were systematically overrun. People's vulnerabilities often worsened after reconstruction. "The key to success" in reconstruction, the agency concluded, "ultimately lies in the participation of the local community—the survivors."[1]

Of course, the idea is appealing. If citizens are welcome to participate in decision making after a disaster, the initiatives that they devise will better respond to their real needs and expectations. Social movements, urban planners, and scientists across North America and Europe had also campaigned for community involvement in projects and policy. The United Nations agencies, the World Bank, and other development banks had similarly encouraged this kind of participation in the 1970s. Now they were followed by a myriad of municipalities, charities, and cooperation agencies worldwide.

Since the publication of the Brundtland Report in 1987, community and citizen participation has become a key component of the sustainable development dream. As the report stated, "participation in decision-making processes by local communities can help them articulate and effectively enforce their common interest."[2] Years later, in 2015, the United Nations' plan for disaster risk reduction, called the Hyogo Framework, argued that the reduction of disasters required "an all-of-society engagement and partnership." This framework explained that sustainable disaster-risk reduction demands "empowerment and inclusive, accessible, and non-discriminatory participation, paying special attention to people disproportionately affected by disasters, especially the poorest."[3]

Today, citizen and community participation has been adopted as a key element in the strategy to mitigate risk and disaster. Participation has also become one of the central pillars in the fight against climate change.[4] One analyst put it bluntly: "Climate adaptation [is] impossible without community participation."[5] For charities, consultants, cooperation agencies, and development banks, participation has become a way of convincing donors that they are not "giving beneficiaries the fish," but rather "teaching them how to fish." For politicians, it serves to show that they keep people's problems and desires in mind, whereas for intellectuals, participation is ideal for realizing social justice. It is also a tool for civil society groups to legitimize their existence. Practitioners and consultants in disaster relief agree that participation is a win-win solution for dealing with complex problems.

This enthusiasm for the participation movement is understandable. In the early decades of the twentieth century, almost all cities were transformed and modernized without considering collateral effects on the most vulnerable. Almost a century later, dams, airports, highways, and other large infrastructure projects are still being built without considering the needs and requirements of indigenous and rural communities. City beautification projects systematically displace poor residents, slum dwellers, and other marginalized groups. Authoritarian governments and ideology-driven regimes continue to implement plans that benefit only a few privileged individuals. International development projects are still executed in poor countries without hearing beneficiaries' requirements or expectations. In general, centralized planning has been at the center of several social injustices in developed and developing countries—and in both communist and capitalist societies alike.

It has thus become expected that, when oppression and authoritarianism linger, we embrace participation as a moral duty and a matter of social justice. For a complex initiative to succeed, a form of dialogue and consensus among the people who promote it—or are affected by it—is key. However, while I am convinced of the merits of dialogue and consensus, there is not enough evidence out there to prove that citizen and community participation is the key to success in reconstruction or disaster-risk management.

In this chapter, I will expose the problems with the implementation of participation. Due to increasing unnatural disasters, these problems are exacerbated to the point where major changes are needed to reevaluate decision making today. In 2017, I moderated an online debate on the effectiveness of participation, including the option to vote on a motion about whether participation was considered crucial for disaster-risk reduction. Hundreds of academics, decision makers, and ordinary citizens commented on the issue and voted on the motion. One of the best scholars in urban issues was in charge of criticizing these participatory approaches,[6] and yet, about 70 percent of those in attendance consistently argued that participation *is* the key to success in disaster reduction and response. Most recognized its limitations but were not ready to take it down from its pedestal.

Acknowledging the root issues and secondary effects of participation is not always easy. But the overall enthusiasm for participatory engagement hardly corresponds to the evidence obtained from research. As a matter of fact, one of the prevailing factors of success in disaster-risk reduction and response is—perhaps paradoxically, given my criticism of master plans in chapter 1—the presence of strong local institutions. This brings us back to the Haitian mayors invited to Louisiana.

BIG PROBLEMS, WEAK INSTITUTIONS

The IAMF was not alone in its objective to create a new institution in the aftermath of the Haitian Goudougoudu. When destruction occurred in

Haiti, most foreign observers, charities, and international agencies noticed that the Haitian state was particularly fragile.[7] Most associated this with an apparent lack of formal institutions, so they pushed to create multicity groups, new ministries, multiministry councils, and other bureaucratic entities. But the real problem in Haiti was not the lack of institutions;[8] it was that those existent entities were largely inoperative.

Institutions in Haiti had significantly eroded over the previous decades. Much like in El Salvador, Honduras, Nicaragua, Guatemala, and other Latin American countries, the main reason for the weakening of the Haitian state was neoliberal policy and market liberalism.[9] During the 1980s and 1990s, Haiti was consistently asked by donors and lenders to reduce deficits as well as the budget devoted to government functions.[10] In order to receive aid, the country was required to "integrate global markets." In terms of the Washington consensus, this meant selling public assets, slashing tariffs on imports, deregulating businesses, and restricting public spending.[11] To pay off the international debt, Haitian governments were asked to constantly amputate government agencies and offices. Experts in Haitian affairs have concluded that these "neoliberal policies, imposed [over] the last three decades, cemented the country's political and economic dependence."[12]

Additionally, as president of the United States, Bill Clinton maintained a series of neoliberal measures that further impoverished Haiti. The United States imposed sanctions that forced Haiti to drop tariffs on food imports and implemented strict reforms that reduced investments in agriculture and food production to focus, instead, on exports. In 1994, Clinton also deployed about twenty thousand troops to Haiti to restore President Jean-Bertrand Aristide, who had been ousted in a coup in 1991.[13] Clinton's economic plan for Haiti was such a disaster that, years later, he actually apologized for his policies.[14] At a hearing before the Senate's Foreign Relations Committee, he admitted that, while "it might have been good for the farmers in Arkansas, it has not worked" in Haiti. He then explained: "It was a mistake that I was a party to. I did that. I have to live every day with the consequences . . . of what I did."[15]

After all those years of imposed policies and American interference, the living conditions in Haiti deteriorated. The country became significantly dependent on imports, entire municipalities broke down, and regional governments nearly disappeared. A few local monopolies ended up controlling the economy—and so, dominated Haiti's political landscape.

The country became largely reliant on aid and charities.[16] In the 1970s, Haiti only imported about 19 percent of the food its citizens ate, as they produced sufficient rice, sugar, poultry, and pork. By 2010, however, after the disaster struck the capital city, 51 percent of food was imported, which led Haiti to become the fourth-largest importer of subsidized U.S. rice in the world and the primary Caribbean importer of foods from the United States.[17]

By 2010, the problem of public administration in Haiti was tied to its lack of administrative capacities—lack of competent technicians and professionals, clerks, secretaries, front-desk personnel, and other staff. As the representatives had expressed in Louisiana, institutions such as the multicity association of mayors already existed—but after years of neoliberal and free-market policy, they had simply become empty shells, with little power, projects, or resources.

From 2015 to 2017, I participated in a project in Haiti funded by the European Union (another experience as an international consultant). The ultimate goal was to create a metropolitan structure for Port-au-Prince—a new version of the multicity association. The idea was, again, virtuous. By 2030, the metropolitan area will have more than 5 million people, so the lack of coordination among neighboring cities will likely restrict the development of efficient transportation, key infrastructure, and public services.[18] This initiative surrounding the metropolitan structure, inevitably, met the same problems I found in Louisiana. Lack of resources had already prevented existing Haitian institutions from functioning. An additional organization, council, or association would just become another burden that cities and ministries could not afford.

And yet, strong institutions are vital in the face of risk. After the earthquake hit Gujarat in 2001, the government reinforced its existing institutions and created a new one—the Gujarat State Disaster Management Authority (see chapter 4). In chapter 7, we will examine a similar phenomenon that took place in Colombia in 1999, when a powerful earthquake killed almost one thousand people. The Colombian government supported existing agencies and formed a competent entity—the Fondo para la reconstrucción del eje cafetero—FOREC. In Cuba, the government has very few resources, but a strong public sector guarantees timely evacuations in the occurrence of disaster. Even postdisaster recovery is well organized. Compare this to some of the other cases we have explored in previous chapters. There were no strong institutions that emerged from the disasters in El Salvador (1998) or Honduras (2001)—nor had they surfaced following the Goudougoudou in 2010. Participatory activities were promoted by charities and agencies in all these cases. But while the results were positive in Gujarat, Colombia, and Cuba, participation has achieved very little in places with weak institutions.

Building and reinforcing supportive institutions is, of course, complex. In 2007, a powerful earthquake hit the Peruvian coast, killing almost 6,500 people and causing major destruction in Ica and other cities.[19] Back then, the government tried to replicate the experience of the Colombian FOREC and created an entity called FORSUR. But unlike its Colombian counterpart, this new organization in Peru was poorly coordinated with existing agencies, and it was politically motivated. It ignored the power and role of local authorities. A 2008 report found that FORSUR was "seen as an entity created in Lima by the central government, which does not take into account the views and perspectives of affected people and their elected local authorities."[20] Ten years later, about eight thousand people in Pisco, Chincha, and Ica, hadn't been able to recover.[21] In 2017, I participated in a working session in Lima, Peru, where I was invited as a consultant to present on the subject of reconstruction. Government officials were trying to reinforce existing institutions to cope with the floods triggered by El Niño. In this instance, they created the Autoridad para la Reconstrucción con Cambios (RCC), an institution that strives to avoid the negative experiences of FORSUR.[22]

As the Peruvian case shows, strong public institutions require effort to maintain. The 2010 disaster in Haiti, and the ones in Central America during the 1980s and 1990s, illustrate just how easily neoliberal policies can erode the existing ones. This is a theme that I initially introduced in chapter 1. Here I will argue that one of the problems with community participation is that it helps to justify actions aimed at reducing the role of the state. It also distracts us from creating the types of institutions that are required to achieve possible change. To be clear, participation is not an offspring of the neoliberal agenda. But today, the rhetoric of participation allows those in power to dilute the importance of reinforcing public institutions. In this regard, it legitimizes the case for decentralizing power and minimizing or eliminating regulations, and as such, disengages public institutions in times of disaster. It also facilitates the transfer of responsibilities to markets and the private sector, which in turn, fortifies private interests.

But before developing this idea, we must first investigate the mechanics of participation. As we shall see, participation is often nothing more than an exercise to legitimize decisions made in advance. To fully appreciate this, we must return to the case I presented in chapter 3 about low-cost housing construction driven by an Irish NGO in South Africa.

COME, PARTICIPATE, AND ACCEPT MY IDEA

When Ms. Mkhize raised her hand, those sitting on the panel were already worried that the meeting had lasted too long. It was already 10 P.M. by then, yet Ms. Mkhize and her neighbors from Bonteheuwel had many things to complain about, so the meeting went on just a little longer. But neither Ms. Mkhize, nor the other black South African women who attended the meeting got what they needed: jobs, better schools, and health care. All they were told was to "come back the following week to paint the houses" that they were to receive "for free."

The project's organizers had anticipated that beneficiaries would work on the construction of their new homes funded by the Miller Foundation, an Irish charity. They were also expected to take part in the dismantling of their existing shacks (the South African term for slum houses built with fragile materials). So the women were invited to weekly meetings to understand the work expected of them. But these meetings were already difficult to accommodate, as many of the women were single mothers with day jobs and domestic chores. It was also common that they looked over elderly family members, which made it difficult to find both babysitters for the young and guardians to care for the old. Occasionally, the women attended meetings with their children and parents. But the meetings were long and boring.

Nonetheless, Ms. Mkhize had pressing concerns. She went on to ask another question: Could she transfer a window from her existing shack to her new house?

The NGO representatives hadn't anticipated that beneficiaries would recover materials from the shacks that they were going to demolish. Nor had they anticipated that those materials could be used in the new homes. They looked at each other as if searching for an answer, but because they wanted to wrap up the meeting, one of the officers agreed. Ms. Mkhize would be allowed to use the window in her new home.

Just when the project organizers started to pack their things, Ms. Mkhize raised her hand again. I could almost see the officers rolling their eyes. Now that she would use her own window in the new house, the Irish charity would "save" on her window. What could she get in return for "spared" materials? Some type of compensation would be fair, she argued.

This question was puzzling to the NGO. If all beneficiaries were given the option of using recycled materials, the quality standards and construction activities would become very difficult to maintain. So the officers preferred to end the meeting without solving this matter. The chairman moved on and asked Ms. Mkhize to raise the issue later—in private.

The outcome was likely familiar to many of the attendees. Ms. Mkhize and thousands of South Africans spend several nights per year in meetings like this one. Because—as almost every government official, NGO officer, and design professional in South Africa will tell you—any successful disaster reduction initiative requires community participation. But scores of South Africans now realize that significant decisions are taken in advance of participatory meetings and audiences.

Members of my research team have joined several community participation sessions across Canada, Colombia, the United States, Vietnam, Cuba, Haiti, Brazil, and South Africa. We have taken part in discussions with frustrated communities aware that participation exercises are, essentially, rhetoric. Here, I paraphrase some of the most common complaints that I have heard from citizens and representatives of vulnerable social groups:[23]

- "They invited us to participate, but the most critical decisions had already been made. We were simply consulted about the color of the façade and other superfluous details."
- "They invited us to the workshops, but when we told them what we actually wanted, they said that it was impossible and that we didn't really understand the complexity of the situation."
- "We were invited to participate, but when we explained what we needed, the engineers used heavy jargon to argue that this was not technically possible. We didn't understand their technical reasons."
- "Project leaders spoke for two hours, explaining the merits of the initiative. Then we only had ten minutes to express our concerns."
- "They asked our opinion and we criticized their idea. But at the end, they did exactly what they had shown us the first time we met."
- "They want us to participate, but the audiences are always scheduled at night when it is almost impossible for us to attend."
- "Since the disaster happened, I have attended countless meetings and workshops. Nothing has changed, and I am tired of those participatory activities."

Now, here is my paraphrasing of the most common complaints I have heard from the other side of the argument—government representatives and officers in charities and agencies. These comments often arise after initiatives meet fierce opposition and criticism:

- "We invite everybody to participate, but there are always the same groups that show up. We know their cause and we know what they want. Yet, they do not represent the variety of realities that exist out there."
- "Social leaders who oppose the initiative are manipulated by the opposition. They are just puppets that pose as civil society."
- "This group only defends its own cause. Members do not participate in a constructive way. They fail to see that our initiative will bring significant advantages for many other people."

- "We invite everybody to express their opinion, but the same people always speak in the meetings. They seize the microphone and intimidate those who are less vocal to express their own ideas."
- "They say that they oppose the initiative. But in reality, they are expressing frustration with policies that have nothing to do with our goal."
- "This social group is reluctant to change because people are afraid of losing the privileges that they enjoy today."
- "Critics of the initiative are simply afraid of change. They don't see the significant long-term effects that can be obtained."

Even the most fervent defenders of, and disciplined consultants in, participation now accept that a climate of frustration, distrust, or open confrontation is all too common in participatory activities. These activities are usually expensive and cause significant delays in implementation. Sometimes people are asked to attend so many meetings that scholars have coined a term for their frustration: "planning fatigue."[24] Many lament that participation activities have become less about decision-making processes and more about public relations. Activists and intellectuals alike know that participation is regularly used as a tool for manipulation.[25] Citizens rightfully perceive that participatory exercises serve to inform the public about the decisions that have already been made. In fact, Latin-Americans increasingly call this the *socialización* of initiatives.

Is planning fatigue, additional costs, frustration, socialización, and long delays the price to pay for democracy? Perhaps. But the price tag often comes with extra costs that are never fully disclosed, which further erodes the trust of public institutions—including the ones that democracy relies on.

THE UNDISCLOSED COST OF PARTICIPATION

In the 1970s, a British architect named John F. C. Turner revealed how slums and shantytowns in developing countries were created. While he wasn't the first academic to study informal settlements, Turner became one of the most influential and celebrated authors in the field. His work revealed how low-income households in the Global South build their homes over long periods of time, harnessing a series of capacities and skills to incorporate basic services, such as water, sanitation, and connectivity to electrical systems. Turner concludes that low-cost housing should not be seen as a product but a *verb*: "As housing action depends on the actors' will," states Turner, "and as the dominant actors in economies of scarcity are the people themselves, they must be free to make decisions that concern them."[26]

Turner's findings appealed to those on the Left and the Right of the political spectrum—albeit for different reasons.[27] For the Left, his ideas highlighted the struggle of the poor and marginalized sectors, and—taken as a whole—proved that those populations are, in fact, strong actors in the economy. They are not lazy or passive recipients of aid, as many radicals on the Right had suggested. To the contrary, they are resourceful agents trying to cope within a hostile environment.[28] For the Right, meanwhile, Turner's findings were exploited to demonstrate that fewer efforts were required to improve the living conditions of the poor, as low-income and vulnerable citizens already have the ability to overcome their most significant challenges. A little bit of support is probably all that they need to unleash their potential and solve their own problems. Of course, this implies that only a minimum investment in housing and services is required for the poor.[29]

The World Bank and other agencies in charge of implementing the Washington consensus, naturally, took notice of this argument surrounding self-sufficiency. The bank looked at Turner's findings and those of other academics in the 1970s who had explored the skills and capacities of low-income residents. Eventually, the bank issued a housing policy for developing countries based on their interpretation of these findings.[30] Aided self-help became a generalized strategy for dealing with the qualitative and quantitative deficits of housing and services in the Global South.[31] Participation was often promoted in the World Bank's Structural Adjustment Programs and the Enabling Markets to Work policy, which implemented neoliberal and free-market approaches in Latin America, Africa, and some Asian countries.[32] Housing programs for the poor would then take advantage of beneficiaries' self-sufficiency. With minimum government intervention or investment, they would be able to build their own housing and infrastructure solutions.

In the meantime, both the United States and Europe's movement of participation was taking hold. Citizens and civil society groups condemned authoritarian solutions and claimed power to influence public policy and projects.[33] The failure of public housing projects, where finished units were given to beneficiaries, was now condemned in the rich world. This resistance was fertile ground for the community participation movement of the 1980s,[34] and so, both local governments and charities were pressed to adopt participatory approaches. They should avoid centralized schemes where residents became the "passive" recipients of aid. Civil society groups had to be empowered to face authoritarian governments and unjust solutions.[35]

When it came to combatting the vulnerabilities of the poor, NGOs from developing countries also remembered the principles revealed by Turner and the self-help policy promoted by the World Bank and the United Nations agencies. In terms of housing, most charities and governments found an easy answer: people can participate in residential initiatives by providing labor. Participation quickly turned into self-construction—or simply put, the provision of cheap labor. In Latin America, the merits of participation were a golden opportunity for a small elite unwilling to invest in the improvement of living conditions for poor and indigenous groups. From Mexico to Argentina, *autoconstrucción* (aided self-help) not only became a housing policy but also an overreaching philosophy for managing the poor.[36]

Millions of people in the Global South were regularly asked to provide labor ("sweat equity" in bureaucratic terms) through construction activities. Charities, governments, and international agencies organized construction brigades where beneficiaries carried cement bags, bricks, and other materials, in exchange for the prospect of receiving a new house and land title. But beneficiaries received minimal training and supervision in these construction initiatives (and maybe a sandwich and soft drink at lunch). In most cases, they had to put up with a certain number of work hours to be eligible for a house, latrine, or water tank.[37]

One of the perverse consequences of this policy was that urban solutions had to rely on a very simple form of construction. This was the only way to involve unskilled laborers. Most self-help residential projects thus resorted to the construction of remarkably simplistic single-story detached homes,[38] an approach that favored suburban-type developments, built in the periphery of cities where land is cheaper. This led to significant increases in urban sprawl and made commuting and transportation a major challenge for low-income residents in most cities of the Global South. The low urban density achieved through self-help programs also challenged the sustainability of the (somewhat limited) infrastructure and services provided in new peripheral neighborhoods.[39]

While public institutions were being eroded in Latin America and Africa in the 1980s, following Reaganomics and Thatcherism, participation was being promoted as a key factor to success. The rhetoric of inclusion was typically used to legitimize neoliberal policies that had two perverse effects in terms of risk.

The first effect involved new and worsening vulnerabilities among the poor. Most low-income families found themselves living in peripheral areas, where lack of infrastructure and low densities made the development of profitable home-based businesses, such as convenience stores and diners, nearly impossible. This took a large toll on uneducated women, who typically relied on home-based businesses to make money. Unable to find skilled jobs, these home-based businesses enabled them to take care of their children and elders. The poor, ultimately, found that they had worked to build a house that had worsened their living conditions.

The second effect encompassed the (sometimes subtle) unveiling of policies aimed at transferring responsibilities from government to companies and individuals, while reducing public spending and limiting the role of the state. As we saw before, neoliberal policy had reduced institutional intervention in the face of natural events, so disasters proliferated in many countries.

The results of participation proved disappointing—even for World Bank standards at that time. In 1987, the bank commissioned a study to evaluate the impact of community participation in more than forty projects across Latin America, Asia, and Africa. The report presented a bleak view of participation: "Bank policies have focused relatively more on project effectiveness, efficiency, and cost sharing than on beneficiary capacity, building, and empowerment." It also concluded that "the indiscriminate use of community participation in all types of projects is unwarranted."[40] In other words, participation had been used to achieve cost-effectiveness at the expense of quality and better solutions to reduce vulnerabilities. In 1996, a second World Bank report found that "bank-financed operations can range from minimal levels of participation (which some do not consider to be participation at all)—such as information sharing and consultation—to intensive participatory mechanisms, such as collaboration in implementation and decision-making."[41]

And yet, the policies of the IMF and the World Bank were frequently backed up by United Nation agencies. In 1997, the United Nations Human Settlement Program (better known as UN-Habitat) set up the policies to achieve "sustainable human settlements." The policy prescribed that "governments should strive to decentralize shelter policies and their administration to subnational and local levels within the national framework, whenever possible and as appropriate." The emphasis on "enabling markets to work" also followed conventional neoliberal ideology. Governments were asked to "avoid inappropriate interventions that could stifle supply or distort demand for housing and services."[42] Instead, they should deregulate markets, sell public institutions, and transfer responsibilities to the private sector.

The expansion of this policy was remarkable. Four years later, UN-Habitat indicated that "there has been an equal consensus in favor of the decentralization of responsibilities to lower levels of government, and, particularly to local authorities."[43] But for intellectuals and scholars in developing countries, participation became confusing. It was too many things at the same time: autoconstrucción, decentralization, empowerment, capacity building, project ownership, and a mechanism of urban management.[44] A UN report on community participation in disaster recovery found that "informing, consulting, involving, collaborating, and empowering communities are the core building blocks of participation."[45] They are also agents of decentralization in the name of sustainability, democratization, and the spreading of decision making among local organizations.

For funding agencies in the Global North participation became a way of showing sensitivity to local problems. "The projects are therefore not only about bricks and mortar," said one Canadian organization, "but increasing the capacity of communities, NGOs, worker unions, local governments, and the private sector to play key roles in the housing process."[46] But again, this was, at best, a cost-reduction approach. In the worst cases, it was a marketing strategy that facilitated the spread of disengagement at the state level.

Some might say that these were the problems of a different era—that participation today should not be confused with its earlier, exploitative version that dominated in the 1970s to the 1990s. Today we tend to associate participation less with self-help and more with design activities. But while it is true that the narrative of participation and implementation mechanisms have evolved, many of the previous problems faced by community and citizen participation remain. In fact, they might even increase as the planet warms up faster and disasters become more frequent.

TWO PROBLEMS: COMMUNITY AND PARTICIPATION

Today, two problems with community participation in both rich and poor countries persist. One is "community," and the other, "participation."

The term "community" is used arbitrarily in reference to many things: a neighborhood, a slum, a group of local NGOs, a bunch of militant leaders, residents of a small town, a workers' union, a group of women, or disadvantaged individuals. The term sounds romantic in most policy documents and academic papers. It is also less problematic than saying "they," "victims," "survivors," or "poor people." But it does not indicate what a group of people really has in common or what differentiates them. It also creates the false illusion that individuals gathered within "the community" share the same objectives and values. In most cases, the term "community" is used to legitimize the involvement of leaders who do not necessarily embody the interests of those being represented. Besides, it masks the current reality of most contemporary societies, where individuals belong to several communities with different—often conflicting—agendas. Urbanities today belong to groups that represent their characteristics in terms of, say, religion, race, sex, politics, leisure preferences, social interactions, and physical conditions. These "communities" have different agendas and voice different concerns in the public realm. By calling a contemporary urbanite a member of *the* community, we render invisible his or her affiliations and expectations. "Communities are not usually coherent entities," write James Lewis and

Ilan Kelman, two researchers in the UK. The authors conclude that "local knowledge expressed through community consultations designed to elicit a coherent 'community view' is sometimes shown to be counterproductive to management decisions, or outright erroneous."[47]

The term "participation" is similarly used to denote a bunch of different things. Today, it is routinely used to describe civil debate and communication, consultation, delegation of activities, partnerships, self-help construction, communal meetings, political decentralization, local referendums, public audiences, focus groups, and collaborative workshops, among others. The mechanisms and objectives that underlie concepts like "living labs" (a trendy term among managers and designers interested in participation) are obscure in the eyes of the general public—as are their value in social and democratic life.

Take ownership, a common directive given by participation and urban consultants. Agencies often claim that, when "community" participates, members gain "ownership" of their initiatives. In reality, this is little more than technocratic jargon. Participation is often a tool to coerce people into accepting predetermined ideas and principles defined prior to community involvement. Think about this: when you have an idea and you see its value, you don't need a participatory exercise to gain "ownership" of it. You just believe in it.

Today, the idea of community participation has been so widely expressed that it no longer means anything significant.[48] Ironically, in times of rampant consumerism and reliance on a capitalist society, participation is the mechanism by which the most powerful often legitimize their objectives, and those who speak loudest are heard. By coupling it with sustainability and resilience, participation shields the most powerful from criticism, while maintaining a prim public image.

FEAR OF STRONG INSTITUTIONS

A more precise version of the UNDRO's motto might as well be: "The key to success ultimately lies in the effectiveness of strong institutions." I reckon that this idea might be taken as support for ideology-based autocratic governments. And yet, the Gujarat case, along with those I will discuss in chapters 7 and 8, show that these types of strong institutions are *not* autocratic. Quite the opposite, they include stakeholders in efficient decision making and create proper governance mechanisms that guarantee control, checks and balances, and the fair distribution of resources.

But most mainstream economists in the West fear that strong institutions become confused with—or simply *become*—oppressive. Dictators in Africa and Latin America, political extremists in the Middle East, despot communist regimes, and world wars are (rightfully) often used to highlight this danger. Besides, the liberal conceptions of the state preferred by Western urbanites often prioritize morally neutral institutions.[49] Surely, neutral, secular institutions focused on universal human rights are key to preserving harmony in cosmopolitan centers, where heterogeneity and multiculturalism prevail. In order to guarantee diversity and prevent the abuses of minority rule, public institutions must avoid religious and ideological biases. This implies a form of moral neutrality from the state,[50] where institutions focus on the arbitration of diverse rights and claims.

We must celebrate the benefits that have brought about secular conceptions of the state in many parts of the world. But we must not fail to see cases in which liberalism is tweaked to legitimize neoliberal doctrines. In a mainstream neoliberal conception of the state, institutions must minimize their role. This implies setting up as little regulation as possible, which gives the "invisible hand" of free markets the opportunity to arbitrate diverse and conflicting demands.[51] And here is where participation becomes useful. The idea that participation is key to success reinforces the image of a neutral negotiation of interests and rights. It permits institutions to referee diverging claims while acting as little as possible in an ideologically-motivated fashion.

Let me be clear, I am not arguing that all charities and professionals that defend virtues of participation have hidden plans to install neoliberal policies. Nor do I believe that they are fervent defenders of radical free-market ideology. In my work, I have found that most charities and professionals promote participation with the genuine intent of improving social and environmental impacts. But a narrative that places participation as an uncontested key to success facilitates the implementation of neoliberal and free-market ideologies. Today, the participation rhetoric serves to dilute (or even mask) the role that strong institutions must play to combat social and environmental injustices. Meanwhile, those who are well-intentioned must confront the difficult task of reconciling the objective of participation with the objectives and needs of the less powerful and less vocal.

BARBECUING THE PARK

"Subsidized housing will attract immigrants," said one of the attendees. "They will then smoke our park with their barbecues," he complained.

The remark provoked outrage. Surely, Canadians had become accustomed to public consultation. The fifty or so local residents who had agreed to attend a public meeting to discuss Montreal's Technopôle Angus project expected controversy, but few anticipated anything quite so radical.

In the days following the meeting, even more citizens condemned this comment—which clearly reflected a "not in my back yard" attitude—as discriminatory. Commentators were furious.[52] The urban developer was baffled. The project presented at the meeting had, at least in theory, all the ingredients for success; it integrated almost every principle of sustainability, proposed innovative systems to reduce energy consumption, and included incentives for bicycle use. The Technopôle Angus project also featured reduced water consumption and generous green areas. It combined workspace, commercial suites, profit-seeking condos, and affordable housing. Low-cost units would be offered to vulnerable families. The plans had been improved through integrated design activities, public participation meetings, and other collaborative strategies. The developer sought sustainability, resilience, innovation, and participation. In other words, the project had all the elements that developers are expected to incorporate in contemporary city-making. What could have prompted the NIMBY reaction?

It might still surprise Canadians to hear discriminatory comments in public audiences. But more frequently, developers in Canada and other countries are encountering NIMBY reactions.[53] As a matter of fact, opposition to projects and policies in times of global warming is becoming a barrier to the reforms we so badly need. In some cases, opposition comes from the bottom. In other instances, it comes from powerful lobbyists and economic elites. Both cases, however, are putting participation to the test, particularly in times when the planet is heating up at rapid pace.

GLOBAL WARMING IN TIMES OF DISTRUST

Climate change is a major tragedy. One of the things that makes it so difficult to resolve is that it is accelerating in times when confidence in democracy continues to fade.[54] Global warming requires strong political action, but action is halted when citizens become discouraged by established political systems and resort to populist politicians.[55]

In postdisaster recovery, participation and lack of trust often follow a vicious cycle. A lack of trust in democratic institutions specifically tends to boost the use of participatory mechanisms. When citizens don't trust elected officers, you might think that they would get involved in correcting their mistakes, setting up goals, and blocking unsound solutions. But there is space for manipulation: participation exercises are too often created to avoid conflict rather than to modify or abandon preestablished initiatives. The consequence is that citizens sooner or later realize that they are being manipulated in city halls, referendums, public audiences, workshops, and through other participatory initiatives. Once they sense that they are manipulated, they further lose trust in the institutions that invite them to participate.

In times of climate change, where difficult decisions must be made, this tendency is likely to continue. After all, the majority of decision makers still don't have answers to the most pressing question raised by participation: What happens when legitimate but conflicting interests emerge?

This is not an abstract question. Conflicts have concrete effects in project and policy planning, or intervention, especially in the face of risk. Take the case of legal responsibility for construction projects. In most countries, architects and engineers are legally responsible for the buildings and infrastructure they design. Their solutions must protect the public and guarantee their safety. Buildings and facilities must not fail when common hazards occur. Failure to accomplish this can lead to prosecution, fines, or even jail.

But these professionals are also increasingly asked to work in an environment of shared decision making, where participatory activities, collaborative workshops, and other techniques are common. Yet the people who encourage citizens to participate do not have any legal liability. They do not get fined or go to jail if the proposed solution fails to protect citizens or results in an accident. Only professionals are held responsible for the blueprints they put their professional stamps on. As such, architects and engineers are often uncomfortable with the involvement of people who have no form of legal liability when it comes to decision making. How can these professionals guarantee honest participation while retaining full liability for their own work?

Take also, for example, the case of urban density. Almost every politician and urban technocrat will tell you that they want to achieve sustainability, which in turn, depends on community participation. They will try to convince you that these objectives are easy to attain simultaneously, despite evidence to the contrary in times of disasters. Quite often, when confronted with a choice, citizens prefer to reduce densities—not increase them.

Here is an exercise: Ask almost anybody in a low-density suburb whether they would accept having a series of ten-story buildings built on their street right now. I am willing to bet that most wouldn't support the idea. Some people might argue that the new towers would create shadows and interrupt their views of the landscape.[56] Others would point out that more residents generate more traffic, noise, and pollution; kids and elders won't be safe anymore. And yet, living in high-density spaces is our best chance at reducing carbon emissions, pollution, urban footprints, and the destruction of both agricultural land and wild landscapes today. Are we going to reach higher densities by asking for people's preferences?

While sustainability and participation are often presented as two sides of the same coin, few decision makers have noticed that sustainability represents a significant challenge for participation. In many cases, they are at odds. Who will defend the rights and freedoms of future generations now? Who, in public audiences and town hall meetings, will petition for restricting the benefits of vulnerable people today, for the sake of those individuals who have yet to be born? Will we deny practical solutions for people suffering now in order to guarantee the rights of others who might suffer one hundred years later? Are we ready to restrict our current freedoms to respect and preserve the rights of future species and humans to come?

These are some of the dilemmas that professionals on the ground are confronting today. City and NGO representatives, for instance, often deduce two main possibilities for achieving higher urban densities in reconstruction projects: One option is to go ahead with participation and increase densities, which would frustrate people who feel like their opinion is being ignored, and ultimately, fuel distrust from the public. The other option is to reduce participation and still go ahead with densification, but this decision would also create frustrated citizens who perceive it as autocratic.

Consultants and other defenders of community participation argue that the baby must not be thrown out with the bathwater. Most recognize the drawbacks of participation,[57] but they argue that the problem lies in implementation—not just the principle itself. If properly conducted, the majority of consultants will contend, participation can overcome these challenges and reach its main objectives.[58]

But even here, the defenders do not all agree. There are those who claim that participation *is* the objective, while others see participation as *a way* of reaching the overall output, such as greenhouse gas reductions. In the first case, participation is the ultimate goal that will empower citizens and consolidate democratic values—for the latter, it is the means to achieve environmentally sound or socially just outcomes. As the architects' legal liability and urban density cases both show, obtaining these objectives at the same time is difficult. Besides, fervent defenders of participation rarely agree on the exact course of action to take. Some point to the methods themselves, while others to the communication of objectives or perhaps the actions that must be tailored to specific conditions—and still, many argue that participation exercises overestimate their objectives, which creates expectations that can hardly be fulfilled.

These are all valid arguments. There is a tendency to generalize practices without considering the specific contexts in which they are implemented. It is also true that there is space for improvement when it comes to implementation. But most consultants and defenders of participation fail to provide practical answers to the most vital questions: Who is a legitimate participant in decision making? What is a legitimate claim? How can we compare conflicting claims? They also fail to recognize that participation is overly compromised. It is not only time to challenge its efficiency but also the effective value that participation provides.

THE VALUE OF PARTICIPATION WHEN MAKING TOUGH DECISIONS

Constrain overconsumption. Charge polluters. Limit urban sprawl. Set up a carbon tax. Raise taxes on gas. Regulate the high-end residential sector to contain increases in home sizes. Restrict the use of carbon-emitting cars and machines. Challenge the financialization of housing. Increase the safety net for the most vulnerable. Strictly regulate Airbnb, Uber, and other multinational platforms. Set up universal "free" health care and education. Ban the use of several industrial products and practices. Block initiatives that generate additional transportation. Expand social housing programs. Impose restrictions on disposable waste in households and across industries. Impede new suburban developments for the rich. Incentivize more people to live in dense urban areas. Limit exploitation of land and ecosystems.

These are, perhaps, the type of decisions needed to contain disaster-risk creation—and they are tough ones to make. In one way or another, they restrict the rights and freedoms that many of us take for granted. In fact, many of these measures are unpopular in the eyes of the public. They must confront powerful lobbyists and mighty economic groups. They all require significant sacrifices to be made by companies and individuals alike—many of which are not just symbolic. They will lead to reductions in jobs, sources of income, and economic means. Many of these measures will also result in reduced tax revenues for the government as well as negative effects for large corporations and small companies.

Today, we must deal with these multifaceted dilemmas. Conflicting objectives and expectations must be considered. Only powerful institutions can make these hard and unpopular choices. These decisions require strong leadership at various levels—from national, to regional, and municipal governments. They require institutions capable of mobilizing less vocal social groups and supporting a vision that does not leave behind the most vulnerable. They demand institutions capable of facing corporate lobbies, market forces, partisan agendas, and the vested interests of the most powerful. Finally, they require institutions ready to defend the rights and freedoms of generations to come, even when this involves restricting those of people who can block action today.

Can current institutions simply serve as referees that overlook conflicting claims, as the liberal Western conception of the state prefers? Hardly. The tough decisions that we need to make now require a moral stance on disaster, risk, and vulnerability creation. We are deceiving ourselves if we think that citizen and community participation will save us from assuming these moral stances. In times of climate change and unnatural disasters, we cannot reduce public institutions' roles to the mediation of individual interests. We require institutions that side with those who are most disadvantaged. We need public leaders who defend their causes and create mechanisms of support for them, applicable now and tomorrow.

Participation requires an understanding of the needs and expectations of those around us. But the success of our action will not only depend on hearing their cause. Success will depend on institutions' capacity to amplify the voices of the most vulnerable and to fight underlying interests and political agendas. It will rely on their strength to take the causes and concerns of indigenous communities, poor and marginalized groups, all the way to policy implementation; restricting the privileges of the rich and powerful and enlarging the benefits for those typically excluded. It will depend on restricting freedoms and rights today so that generations to come can enjoy their own. Participation is a mere component of this ambitious task.

Many believe that this view is too pessimistic, while others argue that we can avoid all the hassle if we simply believe in civilization's capacity to innovate. A great deal, meanwhile, assert that we should keep pushing for what we do best: technological development.

6 Innovation

"We Need Something Really Innovative, They Said"

Someone tried something new: including among the list of lucky numbers a few unlucky *draws. This innovation meant that those who bought those numbered rectangles now had a twofold chance: they might win a sum of money or they might be required to pay a fine—sometimes a considerable one. As one might expect, that small risk (for every thirty "good" numbers there was one ill-omened one) piqued the public's interest.*

—Jorge Luis Borges, "The Lottery in Babylon"

THE PRESIDENTS' DREAMS

"If we do this housing properly," said former U.S. president Bill Clinton in 2011, "it will lead to whole new industries being started in Haiti, creating thousands and thousands of new jobs and permanent housing."[1] Clinton was speaking from Zoranjé, a mostly inhabited flood plain about 7 kilometers north of Port-au-Prince. At his side was Michel Martelly, the musician-turned-politician and newly elected president of Haiti.

Bill Clinton was busy in the aftermath of Haiti's 2010 earthquake. He had been appointed as the United Nations' special envoy for Haiti after serving as UN Special Envoy for Tsunami Recovery following the 2004 Indian Ocean earthquake. In the wake of disaster, the Clintons joined forces with former president George W. Bush to create the Clinton Bush Haiti Fund. The Clinton Foundation led efforts to persuade private companies to invest in Haiti through the Clinton Global Initiative and also mobilized a series of partners to create a reconstruction program called Building Back Better.[2]

The Building Back Better program focused on creating an international design competition for the selection of innovative sheltering solutions for housing reconstruction in Port-au-Prince. Malcolm Reading Consultants, a London-based company, was in charge of running the competition. Partners in this initiative included the World Bank, John McAslan + Partners (an architecture practice based in the UK), Arup (a multinational construction company), and two charities: Architecture for Humanity and Habitat for Humanity International. The Clinton Foundation chose Zoranjé as the location for the Building Back Better exhibition, which involved a display of selected housing prototypes. The exhibition cost more than US$2 million, including US$500,000 from the Clinton Foundation and other contributions made by the Inter-American Development Bank (IDB) and the Deutsche Bank.[3]

Housing solutions to build back better in Haiti were expected to be adaptive, resilient, innovative, and sustainable. About sixteen projects were chosen, with a prototype unit for each of them built in Zoranjé. The idea was that developers and citizens would visit innovative prototypes in the area and select the units that they wanted to purchase. The houses, which ranged from US$21,000 to US$70,000, were all single-story detached units. In Zoranjé, these prototype houses were made of metallic containers, prefab cement boards, recycled materials, and even plastic panels (see figure 6.1).[4] The majority were ill-adapted to vernacular ways of living, with strange forms and distributions. Some were downright funny. It was in the opening of the Building Back Better exhibition that Clinton and Martelly had promised thousands and thousands of jobs and housing.

Universities played a fundamental role in the Building Back Better plan. In 2011, about six of the most well-known professors from the Graduate School of Design at Harvard University and MIT's School of Architecture

FIGURE 6.1 Some of the innovative houses selected for the Building Back Better exhibition.

Source: Gonzalo Lizarralde, 2015.

and Planning gathered to develop a master plan and housing solution for Zoranjé. The project, called the Exemplar Community Pilot Project, was mostly funded by the Clinton Foundation and the Deutsche Bank, but it also received support from Digicel, a Haitian cell phone provider.

After months of work, the group of American scholars produced a lengthy record that, for the most part, shows a thorough understanding of the problem. Overall, the report contains an extensive analysis of the territory, an examination of local materials, primary needs, and several photos and long descriptions of participatory meetings held with members of "the community." Additionally, it devotes several pages to the risks that could result from building in a flood-plain area.

But, by the end, the tone and message of the program shift. These professors of architecture, urban planning, and design suddenly forget about the risks of building in a flood-prone region. And they have a somewhat counterintuitive suggestion: to create a "fab-lab" in Zoranjé! Yes, a high-tech construction laboratory in the middle of a largely inhabited, Haitian flood-plain. According to the report, both MIT and Harvard professors would have a set of "digital fabrication machines located within the community" thanks to their Caribbean fab-lab. This would allow them to "create anything from nick-nacks and toys to furniture, up to full-scale building constructions."

On one of the very last pages of the document, as if nearly forgotten, appears a final recommendation: "A public-private partnership for a sustainable community." This recommendation entails:

- A strategy of private sector involvement and economic development
- Strategic partnerships with Haitian banks for servicing, ownership; assistance and promotion of financial literacy
- Deutsche Bank–designed mortgage finance instruments to allow for an economically integrated community with various housing product options.[5]

After pages and pages of analysis that show real vulnerabilities in Haiti emerges what appears to be the real purpose of the study: a proposal to create the innovation lab, predicated on a partnership between financial companies. Underlying this initiative was likely a deal to sell consulting services with the reconstruction money.

Years later, my colleagues and I conducted research in Zoranjé and encountered a striking reality. The thousands of new jobs and permanent housing projects that Clinton had promised were never created. The Building Back Better program was a major failure. One reason for this was that the houses were too expensive, and Zoranjé lacked basic services and transportation. Another was that very few Haitians were interested in commuting to see the housing exhibition in Zoranjé. Almost no one bought the homes that were promoted.

Zoranjé—a very low density neighborhood in a flood plain—never attracted local businesses or residents. As I explained in chapter 3, the Clinton Foundation left Haiti three years after the housing exhibition's construction. The master plan designed by specialists at MIT and Harvard was never implemented. In fact, these Ivy League professors reduced their travel to Haiti altogether. Over time, most prototype houses in Zoranjé were squatted or donated to low-income Haitians. A few were later vandalized. In all the years I visited Zoranjé, the place consistently looked like a ghost town (see figure 6.2).

FIGURE 6.2 The Building Back Better exhibition in Zoranjé became a ghost town.

Source: Gonzalo Lizarralde, 2015.

This is a good example of how the postdisaster innovation machine works in the Global South. The Building Back Better initiative had all the ingredients of an innovative, sustainable plan: politicians seeking to improve their image and enlarge their business network, foreign consultants selling expensive services, design and construction companies interested in making money fast (often by cutting corners), scholars creating or endorsing innovative ideas, and communities being manipulated through participation exercises and meetings to legitimize the whole thing. As this chapter demonstrates, overconfidence in technological innovation is a key component of the sustainability and resilience fallacies.

CONTAINERS, 3D-PRINTED UNITS, FLOATING HOUSES, AND OTHER POSTDISASTER FANTASIES

I met Graham Saunders when he was in charge of housing and sheltering solutions at the International Federation of Red Cross and Red Crescent Societies. Graham, a generous and intelligent man, was a top officer in what is probably the most influential and dynamic organization of disaster response worldwide—and he was always fairly critical of disaster relief action. He had no problem recognizing the aid industry's shortcomings and previous mistakes, but he remained optimistic that improvements in vulnerability and disaster risk reduction were possible so long as decision makers addressed their errors in an honest and transparent manner.

In 2015, I asked Graham about the impact of technological innovation in times of disaster. I wanted to know the role that he believed technology could occupy in the field. He looked exasperated: "Every week, someone tells me they've invented a new gadget to solve the latest disaster-related problem." But most of them, he added, "miss the point." "Quite frequently, our budget for housing rehabilitation after disasters is US$50 per family!" Can anyone design a $50 solution? he asked.

Graham wanted realistic, innovative solutions, not idealistic concepts. He was referring to a common trend that had dominated over the past few decades: for every disaster, there has been at least one enthusiastic engineer, architect, designer, hobbyist, or part-time innovator proposing a new, high-tech solution. And most of them resort to sustainability rhetoric to sell their products. Here are just some of the most recurrent sustainable innovations in postdisaster reconstruction:

- Houses built with recycled plastics that can be "deployed to disaster-affected areas in just a few days"
- Floating houses, and even cities, that can literally navigate the threats caused by climate change
- Inflatable emergency clinics, houses, and community centers that can "simply" be airlifted to remote areas by helicopter (and powered with solar panels, of course)
- Housing units, clinics, and facilities built with recuperated metallic containers
- Off-the-grid houses and structures built with recycled materials such as tires, bottles, fabric, and paper tubes
- Shelters and domes that can be built with sand bags

Why build your home with regular bricks when we can 3D print it? Why adopt messy local construction technologies when we can use expensive robots and software from our new fab-lab?

Since the 1970s, architects, engineers, consultants, and other inventors have been trying to market these housing solutions after every major disaster. None have succeeded in the long term, at least not consistently, in developing countries. And yet, I am sure that many more will be reinvented in the attempt to strike commercial success.

At first glance, there is nothing wrong with these innovative solutions. For months after a disaster, victims often live in precarious conditions and have no place to work. People suffer, get sick, or even die if they lack proper access to water, sanitation, and shelter. So by facilitating construction and logistics, it might feel like a great part of the problem can be solved!

Except that the real problems are *not* solved. Owning a fully furnished and efficient home is rarely the first priority of slum dwellers in developing countries. After years of talking to—and working with—low-income residents in informal settlements and disaster-affected areas, I have learned that recovering a job or source of income, sending children to school, and maintaining access to both health care and security are more pressing needs. When it comes to shelter and permanent housing, the construction of walls and a roof is not the most difficult task for residents in informal settlements in the Global South. As a matter of fact, many of these residents have built or directed the construction of their own homes. Quite often, the issues that they face are related to their lack of land titles and construction of infrastructure, including water and sewage systems, roads, sidewalks, concrete slabs, and electricity.[6] They also have very limited access to credit and face long and cumbersome procedures to obtain construction permits. These are their main concerns, but they are neither addressed by the technological solutions of postdisaster inventors nor by the flashy reports of "urban and resilience" consultants.

Most of the time, technological innovations contribute to divert investments that should be aimed at the most vulnerable. Donors and agencies alike harness technology to make strategic investments to a reduced number of (predominantly foreign) companies. The vast majority of American aid after the 2010 earthquake hit Haiti went to U.S. consultants and companies. It is estimated that "the Government of Haiti received just 1 percent of humanitarian aid, and between 15 and 20 percent of longer-term relief aid."[7] This was also the case after Hurricane Maria struck Puerto Rico in 2017. According to the Center for the New Economy, an independent think tank, about 90 percent of reconstruction funds received in the first year after Maria were granted through contracts with U.S.-based companies.[8]

Technological innovation also accelerates the commodification of housing.[9] By focusing on the acquisition of "new" solutions and high-tech innovations, reconstruction programs transform disaster victims into consumers.

And even when inventors and consultants try to tackle real local problems, implementation becomes a huge barrier. Take the example of credit. As I explain in my previous book, *The Invisible Houses: Rethinking and Designing Low-Cost Housing in Developing Countries*, slum dwellers and poor citizens are not able to fulfill the basic requirements needed to be eligible for traditional financial services. Mortgages are rarely accessible to the poorest families in developing countries, as banks would rather not lend to citizens lacking a permanent job or legal address, which is often the case for the most vulnerable in the Global South. Besides, slum dwellers do not typically have a land title, meaning that their property cannot be used as collateral for credit—not to mention that procedures to legalize their land can take several years. Building infrastructure and implementing financial solutions, similarly, take a long time, so the advantage of rapid construction is lost in the overall housing development process.

There are also several technical problems surrounding these high-tech solutions. First, they fail to recognize the importance of tradition and rooted forms of living. A house is not simply an object; it is a process and a representation of culture, lifestyle, and belief systems.[10] Cultural traits are important to people, even (or perhaps more so) when they are affected by a disaster.[11] They are crucial for rural residents, as much as for slum dwellers and urbanites, in general. Ignoring the rituals and processes involved in housing development is a mistake—so is ignoring the traditional materials, technologies, forms, and layouts of homes and settlements for the sake of "speed" or "efficiency."[12] In chapter 2, we saw that low-income residents sometimes abandon houses and infrastructure solutions that have been provided to them for free. One common reason for this is that the "innovative solutions" used to build their homes concentrate on the technical aspects of construction yet fail to address the traditions, desires, expectations, and values that are most vital to residents.

Another problem with these innovative ideas is that they ignore contextual characteristics. The objective of most high-tech innovations is to find a standardized solution to facilitate construction and streamline transportation and assembly. But this is precisely the problem. Any competent architect will tell you that housing design must respond to the climate, winds, topography, soil conditions, and other characteristics respective to the areas they are built in. Standardized solutions respond poorly to different combinations of these variables. That's why most prefab constructions are either too cold, too hot, poorly ventilated, or badly positioned on the land.

A third problem is that most high-tech or prefab solutions are expensive, particularly when applied in poor countries. Designers and other experts in prefabrication often tell me that their proposed shelter will "only" cost seven thousand dollars. Most of them, either deliberately or unintentionally, omit that the cost of the shelter they cite only represents a small fraction of the total solution. Land, construction permits, infrastructure, a slab on the ground, transportation of materials, assembly, connections to city services, and other activities are repeatedly excluded from the price that inventors use to promote their "new" ideas.

The fourth problem is that innovators and foreign consultants often count on building postdisaster housing solutions on the periphery of city centers, where large parcels of land are cheaply available. Their motivation for doing this is that the supposed cost-efficiency of their solutions depends on mass construction (to reach economies of scale). But jobs and services are typically located in densely populated areas, where land is rare and more expensive. As the Zoranjé case exemplifies, master-planned residential projects on the cities' outskirts are rarely attractive to low-income or disaster affected people.

LEAP-FROGGING INTO THE FUTURE

The overarching argument of high-tech sustainable developers and consultants is that poor communities in developing countries can leap-frog from a preindustrial level of development to a high-tech stage. The idea is that "communities" in developing countries can jump directly to the newest sustainable solutions that pollute less and increase efficiency. By adopting "green" technologies, they can solve their current problems while avoiding the issues of early industrialization experienced by rich countries.

Holding out hope that innovation will produce sustainable construction in the Global South is nothing new. Ever since the Brundtland Report of 1987, the dream of sustainable development in poor countries has been cemented in the production of new technologies. "The capacity for technological innovation," argues the Brundtland Report, "needs to be greatly enhanced in developing countries so that they can respond more effectively to the challenges of sustainable development."

The transfer of technology from the North to the South was a primary objective for Brundtland and her colleagues. "Not enough is being done to adapt recent innovations in materials, technology, energy conservation, information technology, and biotechnology to the needs of developing countries," the report outlines. For Brundtland, the North had many innovative solutions at hand. It was just a matter of convincing poor countries to buy them. In terms of urban challenges, the report argues that, "most industrial countries have the means and resources to tackle inner-city decay and linked economic decline. Indeed, many have succeeded in reversing these trends through enlightened policies, cooperation between the public and private sectors, and significant investments in personnel, institutions, and technological innovation."[13]

More than three decades later, the argument is still widely endorsed by international consulting agencies and even some academics. Steven Pinker, the Harvard professor I introduced in chapter 5, writes, "The less developed countries have a lot of catching up to do, and they can grow at higher rates as they adopt the richer countries' best practices."[14] Pinker and other defenders of technology transfer to the South ignore three key factors: One is that most high-tech solutions increase the dependency of poor countries on rich ones. Another is that most technologies are introduced in the South before adequately testing them. And finally, many "best practices" in the North are poor solutions to the problems in the South.

In chapter 7, we will see that there are solutions to many of the problems posed by postdisaster housing reconstruction. But rather than a new technological gadget, they require political will, engaged local leaders, strong support from public institutions, long-term engagement and collaboration with multiple stakeholders, expensive and long administrative efforts, a group of committed professionals, and ongoing patience. Oh, and take note that none of these produce large profits.

Technological innovations don't just proliferate after disasters in the Global South. They are also perceived by consulting agencies, politicians, and corporations as the most promising way to rebuild in richer countries.

SMART PIGS AND GUINEA PIGS

L'hanno presa i porci	The pigs took it
la mia Aquila e i suoi	my L'Aquila and the
castelli intorno.	castles around him.

In April 2009, an earthquake struck the historic Italian city of L'Aquila and its numerous surrounding villages. The disaster killed 308 people, while leaving 1,500 injured and 67,500 homeless. A university town in the mountains of central Italy, L'Aquila had about seventy-three thousand inhabitants when the disaster occurred. Corruption and political incompetence would come to characterize this event. Under the leadership of Prime Minister Silvio Berlusconi and his cabinet, disaster response focused on monopolizing leadership, shunning local efforts, and favoring those controlled by the interests of the national government.[15] Berlusconi used the reconstruction to increase his popularity and exercise his noxious form of leadership.[16] The insults and dishonesty that dominated the government's reaction to the disaster were captured in a documentary by the Italian film director, Sabina Guzzanti, called *Draquila* (a witty mix of "Aquila" and "Dracula"). In *Draquila*, you can see Berlusconi playing political games and making jokes while people suffer from the destruction.

One of the most controversial measures in the early phases was to prevent residents from returning to their homes in the historic center—a measure that was imposed with abusive authority. Political leaders rushed to start a program to build temporary shelters and permanent units in the towns' peripheries. Very soon, the government announced a program to create nineteen new settlements of earthquake-proof, sustainable, eco-housing complexes, called Complessi Antisismici Sostenibili ed Ecocompatibili, or Progetto CASE. Eventually, about fifteen thousand Aquilani (local residents) were moved to the innovative eco-buildings, and 8,500 more to temporary units. Several solutions focused on standardization and prefabrication.[17] In the CASE buildings, residents found fully-furnished apartments, many of which included "furniture, cutlery, bed linen, flower vases, chairs, and tables [that] belonged to the Italian state."[18] But there were few services, stores, and markets, as well as poor public space and transportation in the area.

In 2010, I attended an opening of *Draquila* at a disasters event in Switzerland. A woman from L'Aquila addressed the group of scholars and disaster aid officers. She knew that Berlusconi's government was already offering furnished apartment units in *CASE* and was offended by the idea. With tears in her eyes, she explained that she was a victim of this disaster—certainly not a rich person—but not someone who needed government charity. "I don't need the government's furniture or pity. I need the opportunity to rebuild my old family home in the city center and recover the furniture and goods my family has had for decades," she said.

She was not alone in her despair. Many survivors called the government's response "the second earthquake" (*il secondo terremoto*).[19] In the beginning, the Italian state was "an agent of hope," but as Forino and Carnelli put it, it became "a source of hopelessness and uncertainty, fostering a sense of crisis."[20] David Alexander, an international expert in reconstruction, says that the response in L'Aquila opened up a "Pandora's box of unwelcome consequences, including economic stagnation, stalled reconstruction, alienation of the local population, fiscal deprivation, and corruption."[21]

Homes in CASE were about 1.6 times more expensive than in other similar buildings. Agricultural land and historic landscapes were lost for the construction of the new buildings. Even more, the new towns caused a dramatic dispersion of disaster victims, which compromised social cohesion and valuable networks. The massive internal displacement caused by the CASE settlements became a major tragedy. Residents and scholars deplored the lack of services and infrastructure available in these new settlements, which led hundreds of "beneficiaries" wanting to relocate.[22] Researchers also pointed out that the new CASE settlements reiterated "the dysfunctions and inefficiencies of urban sprawl, in which the socio-spatial coherence is broken and stimulates the use of the private car."[23] It is no wonder so many of the CASE units were eventually abandoned.[24]

Nine years after attending the premiere of *Draquila*, I met Isabella Tomassi, a native Aquilana. Isabella is a young and smart scholar, as well as a great poet. The day I met her at a disaster-reconstruction event, she read the poem I transcribed above, which she wrote, exasperated by the injustices that have surrounded the reconstruction of her hometown. But by this time, she was less worried by the prefab buildings of CASE than by the Smart City innovations being implemented in L'Aquila.

For years, Isabella has been following the Smart City program underway in the Abruzzo region, where L'Aquila is located. The program includes innovations in mobility, energy, and new industries, as well as hyperconnected services, technology-enhanced interaction between government and citizens, and what technocrats and engineers now call a "5G experimental ecosystem."[25]

The Smart City program in L'Aquila is based on a blueprint written in 2013 by the OECD to rebuild the city. The OECD argued that L'Aquila should "aim to become a laboratory of innovation and promote demonstration effects for the region and other cities." As prescribed by the OECD, the program is based on a partnership between Italian authorities, businesspeople, and academics—all convinced that L'Aquila could become the next European hub of innovation and new technologies.[26]

One of the earliest measures was to create the Gran Sasso Science Institute (GSSI), a center for the study and advancement of new technologies. "L'Aquila has become an open-air laboratory of innovation," Isabella told me in 2019. "Scholars, technocrats, and politicians are playing technology with my hometown." She is not exaggerating. An academic article published in 2018 was unapologetic about describing the old city of L'Aquila and its surrounding historic villages as a blank slate. L'Aquila "has the specificity of being a 'living lab' as it is currently passing through the reconstruction process following the earthquake of 2009, thus offering a 'green field' scenario to the scientific community."[27] After the earthquake, scholars from GSSI and the University of L'Aquila were some of the main drivers of innovation in the city.[28] Problems with traffic? Try "infostructure," they proposed, while asking for open-data and open-source software to optimize mobility in L'Aquila.[29] Issues monitoring structures? Try "information and communication technologies," argued another group.[30]

Unsurprisingly, private companies have shown an interest in the real-life experimentation taking place in L'Aquila. IS Clean Air Italia, a company that specializes in technology to reduce air pollution, calls L'Aquila "an open-air laboratory for new urban solutions."[31] According to one group of researchers, "the structural monitoring, realized through a pervasive use of Information and Communication Technologies (ICT), can be more easily pursued during reconstruction in . . . areas severely affected by the earthquake as in the case of the city of L'Aquila."[32] After several experiments, L'Aquila went on to win a Smart Communities Award in 2015.[33]

For Isabella, technology has created a unique sort of convergence in the Italian society. The dream of a hyperconnected territory in Abruzzo,

Isabella argues, "brings the Left and the Right together, around an idea that benefits companies and scholars—yet not the people affected by the disaster." Many, including Isabella herself, are worried that the new technologies will be difficult for the most vulnerable to adopt, including immigrants, the mentally ill, and older people who constitute about 21 percent of the population in L'Aquila.[34] In 2019, another woman from the region told me that technology is being implemented without knowing why. "The city center is now empty. You have the technology but you don't have people," she said. The main reason is that, by relocating residents to the CASE buildings in the periphery, town centers have become less attractive. "There has been a series of inventions in the reconstruction process, starting with CASE," this Aquilana told me. "The Smart City program is simply the latest in a chain of technical experiments."

Isabella is also increasingly worried about the language used in the Smart City program. "Authorities use all the popular keywords" she contends, "the decorative terms are there: sustainability, creativity, innovation, security, connectivity, participation, etc." But the politicians and technology enthusiasts "are not really interested in social justice, equity, disaster prevention, or preserving the historic landscape." Isabella believes that behind this emulation of the Silicon Valley narrative lies the reproduction of domination and control through technology. "Data permits control of people and space," she deplores. For Isabella, the irony is that "technology does not create local jobs. Innovators are replacing old industries with new ones, and the winners in this game are not from Abruzzo."

A study published in 2020 found that, in L'Aquila, "national, international, and local redevelopment plans have acted as post-disaster 'fantasy documents.' . . . Their hastily crafted projections were based on the overestimation of capabilities of the local innovation system (university-economy), ignoring path dependence and the unfavourable socio-spatial characteristics." The author concludes that the potential for innovation in the context of the "knowledge city" was exaggerated, and the "post-earthquake reality of deteriorated spatial and socio-economic properties was downsized."[35]

L'Aquila is not alone in turning to technology after disaster. The dream of technology continues to proliferate elsewhere.

A CLEAN SHEET OF PAPER

In 2008, I attended a reconstruction conference in Christchurch, New Zealand, where I had the privilege of staying in the city center. I enjoyed nice cafés, restaurants, charming shops, and took many long walks appreciating the local historic architecture, vibrant pedestrian passages, and public squares. Only three years later, most buildings in the central district had disappeared; the city center almost erased. Two earthquakes are partially to blame for this. The bulldozers that followed did the rest.

The first earthquake occurred in September 2010 and the second in February 2011. At least 185 people died by the second event, with 60 percent of the city's five thousand businesses badly affected. Approximately fifty thousand employees had to be displaced. More than 150,000 homes, or three quarters of Christchurch's housing stock, suffered damages,[36] and about 1,100 buildings were demolished in the city center.

One of the government's early measures was to create the Canterbury Earthquake Recovery Authority (CERA). Gerry Brownlee was in charge and became the "reconstruction minister." The government issued a planning document called the "Central City Blueprint," which served to inspire confidence among investors and companies. "I anticipate a light, airy, college-campus style feel for the home of numerous innovative Christchurch companies and public sector agencies," Brownlee explained.[37]

Two years after the disaster, the City of Christchurch applied to take part in the Rockefeller Foundation's 100 Resilient Cities Centennial Challenge. Authorities also asked IBM to provide advice for rebuilding the city. The idea was that IBM could recommend how "to make Christchurch a great place to live, work, play, and do business."[38] Months later, the government launched an urban development strategy with a comprehensive vision: "By the year 2041, Greater Christchurch will have a vibrant inner city, and suburban centres surrounded by thriving rural communities and towns connected by efficient and sustainable infrastructure."[39]

"Technology was always present in the reconstruction process in Christchurch," said Yona Jébrak, a researcher who has studied Christchurch for years. "Technology was seen as a way to gather information for the Blueprint, connect with people, and communicate ideas." Christchurch put forth an ambitious agenda to become a Smart City. Some of the preliminary initiatives, called Sensing City, focused on installing sensors in buildings and infrastructure all over the place. This plan promised that businesses would "thrive, supported by a wide range of attractive facilities and opportunities."[40] In 2013, Brownlee deemed Sensing City a "world-leading project" capable of transforming Christchurch into a "smart city of the future."[41]

The recipe in Christchurch was similar to that in L'Aquila: a partnership between government and private companies; rebranding the city through a powerful marketing strategy; devising expensive contracts for foreign urban consultants; including universities to legitimize change; and controlling vast amounts of data. Everything, of course, was packaged as "eco"—sustainable and resilient change.

Like the companies in L'Aquila, IBM was unapologetic about its lack of respect for Christchurch's past. "Out of the tragedy," reads the IBM report, "has emerged a unique opportunity: not simply a 'once-in-a-lifetime' opportunity but a 'once' opportunity to revisit an entire modern city with a clean sheet of paper."[42]

The objective to attract a creative class through university programs and "cool" businesses mirrored what had happened in L'Aquila. And much like the scholars in L'Aquila, these academics in New Zealand took advantage of the Christchurch disaster. A group of them wrote: "Imagine you had the chance to rebuild a city. Imagine you were able to get continuous flows of data about all sorts of things. . . . This is the idea behind Sensing City."[43]

If Berlusconi was telling bad jokes during the disaster in L'Aquila, the humor in Christchurch was coopted by the company in charge of rebranding the city. Some of the slogans of the program include: "Fun is our fuel," "Be brave, take risks," "Perfect is the enemy of good," and "Innovation moves too quickly to focus on perfect—we get the job done."[44]

Some of IBM's recommendations were so disconnected from victims' problems that it is hard to imagine how they justified proposing them in the aftermath of major destruction. IBM encouraged the city to "Host a 'Geeks on a Plane' program" and "invite entrepreneurs and start-ups from Silicon Valley and other innovation hubs to talk about new markets."[45] IBM appears to be more interested in seeing American geeks opening businesses in Christchurch rather than locals offering public services, social assistance, and financial support to disaster victims.

The urban rebranding strategy is almost comical, and yet the city took the recommendations seriously. It is estimated that by 2018, the Christchurch city council had already established twenty partnerships with private companies and groups, with the vision of fulfilling its Smart City agenda.[46]

In 2017, Brownlee stepped down as reconstruction minister amid national controversy.[47] An auditor-general report found several mistakes and delays in CERA's work. According to the report, CERA "failed to communicate and engage well with the community"[48]—a harsh assessment of the reconstruction process that supposedly uses technology to bring decision making closer to the public. The report also found tensions between CERA and the city council. In 2017, the auditor-general observed that "many people in the region are still facing challenges in their daily lives. Most households have yet to settle their insurance claims or complete repairs to their homes."[49] A 2017 study on reconstruction efforts in Christchurch found that "the highly contested, but more often, misunderstood terms used to construct the narrative of smart cities leads to misaligned expectations, processes, and outcomes." The authors concluded that "being a part of global smart city initiatives, such as IBM Smart Cities Challenge or 100 Resilient Cities Network, Christchurch might seem to be rather progressive, following the modern trends in city governance. However, the ongoing tension between what a city needs, and what corporate or private sector interests suggest cities need, is apparent."[50]

Over the past few years, journalists have reported several implementation difficulties in the Smart City program. Various Sensing City initiatives, for instance, have been abandoned.[51] This is not totally surprising, as innovation enthusiasts are often better at proposing ideas than following them to completion in postdisaster contexts. This is because once the high-tech inventors have sold the idea or obtained the start-up grant, they begin to explore new markets and sales. Consultants, similarly, sell their services and move along to other places. Going to another disaster area is easier for IBM's geeks and urban-resilience consultants compared to dealing with the hassle of real-life implementation.

Despite implementation troubles, it is likely too late to stop the rebranding campaign in Christchurch. A 2016 report found that "progress for the Christchurch projects has been significant, with major milestones being reached." The report noted some delays but assured that "most of the projects are ahead of schedule, with some producing analytics already."[52] Additional investments in sensors, data collection, and high-tech waste management were announced in 2017.[53]

THE BOTTOM ROOM AND THE PHONY CRUSADE TO SAVE THE WORLD

For my Italian colleague Isabella, this frenzy of technology and innovation in the face of destruction hides a more cynic reality. "Technology that collects and uses vast amounts of data about people," she told me, has become "the new panopticon, where those in power hope to have a bottom room to control space and society."

"Is this just a natural evolution of the economic system?" I asked her. Hardly so, she replied. "Many of those who drafted the original recommendations for L'Aquila, were later employed in universities and public entities. This is barely a coincidence."

Isabella has many reasons to be worried. Innovation in times of disaster rarely benefits the local population. Instead, it rewards a battery of external consultants mandated to evaluate the situation, produce recommendations, implement ideas, provide technical support, monitor change, and train local technocrats. The dream of the Smart City requires an alliance among consulting agencies, technology companies, investors, and politicians. And this dream is increasingly packaged as a solution to rebuild after disasters.

After all, disasters are often seen as ideal moments to start from scratch, as decision makers and consultants find less opposition to change. Permission is granted to test and explore the changes that will help authorities consolidate power and companies make fast profits.

Besides, disasters are excellent marketing platforms for both politicians and consulting or technology firms. Berlusconi used the disaster in L'Aquila to improve his popularity. Bill Clinton used the disaster in Haiti to promote his image as a humanitarian. Tesla's CEO, Elon Musk, capitalized on postdisaster Puerto Rico for free advertising.[54] Judging from L'Aquila and Christchurch, it seems that all it takes to get international attention are a few solar panels, new apps, and free Wi-Fi in disaster-affected zones.

When cities and services are controlled by IBM, Cisco, Siemens, Hewlett Packard, and Rockefeller-related companies, it is not the interests of the most vulnerable that will come first but the profit. It is worth asking: What the future would look like if fully planned by tech corporations, consulting agencies, and international firms?

When the COVID-19 pandemic hit Canada in 2020, people in Toronto got a glimpse of how smart, green cities designed by high-tech leaders and consultants work—or rather, how they fail to respond to the needs and expectations of the most vulnerable in regular conditions and in situations of crisis.

A 2019 report painted a sad portrait of the housing market in Toronto. It projected worsening conditions among the most vulnerable, notably "low-income households; seniors with multiple health conditions and fixed incomes; lone parent families, households receiving social assistance, and immigrants." The study found lack of rental housing, areas of population decline, long social housing waiting lists, and other ills. It highlighted, for instance, that about 148,000 additional bedrooms would be required to eliminate overcrowding in the rental market and that there were close to one hundred thousand households waiting for social housing. "In the absence of government intervention and action across the housing continuum," the report concludes, "Toronto's low- and moderate-income households will face a grim housing situation."[55]

The situation was already alarming when the COVID-19 crisis hit both the Canadian and global economy.

Before the pandemic, Sidewalk Labs, a start-up owned by Google's parent company, Alphabet, was on a crusade to build "the most innovative district in the whole world."[56] This hypermodern, model neighborhood was to be built on 16 acres of land in Toronto's eastern waterfront. At the time, Dan Doctoroff, then CEO of Sidewalk Labs, said in several interviews that the whole concept was about "creating affordable housing in the city centre." He often added, as though mimicking Clinton in Zoranjé, that it was also about producing "an economic engine, where Toronto becomes the hub of global urban innovation and generates tens of thousands of jobs."

Doctoroff is supposed to be an experienced leader in urban and reconstruction matters. As the former CEO of Bloomberg L. P. and previous deputy mayor of New York City under Michael Bloomberg's administration, he was active in the reconstruction of the World Trade Center site after the attacks in 2001. Most recently, Doctoroff was busy explaining his new vision for Toronto. In Winter 2019, he announced Sidewalk Labs' Master Innovation and Development Plan. According to this plan, hundreds of affordable units were to be built in thirty story–high buildings. Technology in all its forms laid at the core of the project, which promised new eco-materials and green construction methods, as well as sophisticated communication, information, and transportation systems. The ultimate goal was to create connections among data, buildings, and infrastructure.

But not everyone was convinced by Sidewalk Labs' green approach. Canadians wondered what might happen when high-tech corporations start using personal data collected from urban infrastructure and buildings. Like in L'Aquila and Christchurch, some Canadian activists believed that this was "a corporate takeover of what should be a democratic, government-led process."[57] To dissipate criticism, Sidewalk Labs assured that it had partnered with several public entities that represent citizens' interests. And, of course, their team conducted several "consultations and participatory activities" with local social groups.

In terms of housing, Sidewalk Labs resorted to some "classic innovations": green roofs, as well as flexible, adaptable, and prefabricated housing units that can be inserted in a modular structure. It also proposed clean and rapid construction processes in addition to integrated open spaces that were to be built through modular solutions. (While these ideas might sound original, they have been in architects' minds for more than sixty years.[58]) But most Canadians believed that the project's purpose was not to build faster, more efficient, and less polluting buildings but to use artificial intelligence to collect data about their behaviors—under the pretence of optimizing traffic, logistics, and infrastructure. With the arrival of high-tech urban solutions, some Canadians feared another form of housing commodification, while others thought this was another case of corporate urbanism.

In the end, nothing came of it. The COVID-19 crisis and the subsequent lockdowns brought a significant impact not only in the local economy but also in global markets. A few weeks after Canadians were forced into lockdown in response to the pandemic, Sidewalk Labs abruptly announced that it had abandoned the Smart City project in Toronto.

It is understandable that American companies such as Alphabet were severely hit by the economic effects of the pandemic, and that COVID-19 brought uncertainly about people's behavior in the residential market. But the fact that Sidewalk Labs pulled the plug during the coronavirus crisis demonstrates one of the main problems of corporate urbanism: urban development led by private interests depends on profit, which will always be prioritized above helping the needy.

To be fair, economic downturns affect regular public investment, too. So they also affect urban projects led by municipalities and central governments and built with public resources. But cities and governments are not expected to rely on short- or medium-term profits to decide whether or not to develop solutions for the most vulnerable. Governments are rather expected to be able to act in crises and attend to those who need help, such as low-income families who require affordable housing. Corporations, instead, can easily withdraw from this responsibility whenever the economy goes sour. They are rarely willing to accept deficits to help the most vulnerable.

This tension between corporate profit and helping the poor emerged in Toronto with the Smart City project. Some Torontonians argued that the objective was never about solving Toronto's problems, and that it is possible that "the company was just looking for an excuse to pull the plug on the project."[59] A critic of the project concluded, "Sidewalk Toronto will go down in history as one of the more disturbing planned experiments in surveillance capitalism."[60] Canadians learned that relying on corporations to solve a city's housing problems is like relying on McDonald's to reduce obesity.

American journalist Anand Giridharas wrote that "today's titans of tech and finance want to solve the world's problems, so long as the solutions never, ever threaten their own wealth and power." In his article, published in the *Guardian*, he refers to this agenda as "the new elite's phony crusade to save the world."[61] While disasters pave the road for this crusade, it is not always easy to sell to an informed public. That's where university professors and labs come in. The involvement of academia serves to legitimize radical change and innovation. New ideas are easier to implement if a professor or university report backs them up.

I believe that the majority of my tech-enthusiast colleagues in academia do not have sinister plans in mind. But due to the prospect of lofty grants and prestige, most fab-labs, living-labs, or other university tech-labs, find themselves playing the postdisaster innovation game of politicians and corporations. There is nothing wrong with scholars partnering with governments, consultants, and private companies to develop housing and infrastructure—this is the very essence of technological advancement. But marketing untested ideas in disaster-affected areas and promoting them as "solutions," while making very little effort to tackle the real problems—or to even stay on the ground long enough to guarantee successful implementation—is dishonest.

Disaster-affected cities in rich countries are sometimes viewed as ideal "blank pages" to gain profits and consolidate power, all while obtaining free publicity and scientific endorsements. But as the Toronto case shows, technology is also increasingly portrayed as a way to achieve constant reductions in carbon emissions . . . or should I say "the drawdown"?

TECHNOLOGY WILL SAVE US (FROM THE IMPACTS OF TECHNOLOGY?)

The Most Comprehensive Plan Ever Proposed to Reverse Global Warming. That is the subtitle of *Drawdown*, a 2017 bestseller edited by American environmentalist and entrepreneur, Paul Hawken. Hawken and his colleagues from Project Drawdown, a think tank founded in 2014, are decidedly ambitious. They want us to associate the term "drawdown" with "the point in the future when levels of greenhouse gases in the atmosphere stop climbing and then start to steadily decline, ultimately reversing global warming."[62]

In the *Drawdown* book, website, and TED Talks, the team identifies and ranks eighty of the most effective actions we can take to reduce global warming. The tone of the book is, of course, as inspiring and optimistic as any popular TED Talk. Listed are the types of solutions "that exist today" and that can be "easily" scaled up almost anywhere. Ranked first is controlling leakages from air conditioning machines and refrigerants through better processes to replace, recycle, and dispose of rundown refrigerants. Most refrigerants, you learn in *Drawdown*, use hydrofluorocarbons (HFCs), which "have a 1,000 to 9,000 times greater capacity to warm the atmosphere than carbon dioxide."

Drawdown operates like an encyclopedia of climate change solutions to enhance every facet of our society—from transportation, materials, and land use, to electricity generation, food production, and the construction of buildings and cities. There are also ideas for improving living conditions for women and girls, with an emphasis on the importance of having fewer children: "Securing women's right to voluntary, high-quality family planning around the world," *Drawdown* finds, would save us about 51 gigatons of CO_2—so add it as number 7 to your to-do list.

As you might have guessed, not all the solutions outlined in *Drawdown* derive from sophisticated technology. But the above calculations of carbon savings, which openly are, have already sparked criticism. In the journal *Science*, Faye Duchin, a professor at the U.S. Rensselaer Polytechnic Institute, writes, "The reader is told several times that the Drawdown project is based on measurement, mathematics, and rigorous modeling by scientists and researchers, but this is a case that remains to be made."[63] Duchin finds holes in the methodology as well as the tendency to ignore profound differences in the implications that these measures can have in different regions. Conclusions regarding the benefits of producing green buildings, for instance, stem from broad generalizations and rapid assumptions. They notably fail to account for the energy that is required to manufacture, transport, and implement the new machines, insulation technologies, and other green gadgets contemplated in the study.

This said, *Drawdown* should not be judged like a specialized study on every possible technological solution. If it manages to inspire a new generation of environmentalists or to convince us to change our consumption patterns and behaviors, then plenty has already been achieved. In some respects, its optimism must be celebrated. But, despite its own disclaimers, *Drawdown*'s conviction that technology will solve global warming is overconfident. In this regard, it does not differ from other publications that have shown exaggerated enthusiasm for the impacts of solar panels, wind power, electric cars, biofuels, or systems that capture carbon to prevent future disasters.

Ever since the Brundtland Report was published in 1987, technology has been viewed as a solution to atmospheric pollution and other environmental issues. "Technologies are needed that produce 'social goods,' such as improved air quality or increased product life," the report anticipated. And yet, in the 2000s, scholars already alerted us about the limits of technological advances to reduce energy consumption and greenhouse gas emissions. In 2007, Richard Eckaus of MIT and Ian Sue Wing of Boston University found that "in spite of increasing energy prices, technological change has not been responsible for much reduction in energy use, and it may have had the reverse effect."[64]

In terms of recent commercial books, only a limited number of intellectuals have dared to criticize the main tools of sustainability, such as renewables and electric cars. One of them is Ozzie Zehner, an American writer and visiting professor at University of California-Berkeley. Zehner is the author of *Green Illusions: The Dirty Secrets of Clean Energy and The Future Of Environmentalism*. In his book, Zehner labels many of these recent innovations "the energy spectacles." For him, such technological solutions actually narrow our focus, "misdirect our attention . . . side-track our most noble intentions, [and] limit the very questions we think to ask."[65]

Zehner's supporters believe that "the vision of the future that looks like the present but with 'clean' power is as much a lie as anything the climate deniers say."[66] Zehner points to electric cars, for example, which he finds to be problematic, in part due to the pollution produced by batteries. Subsidies for electric cars, he finds, only further support the type of car-dependent infrastructure and lifestyle that has produced the environmental disaster we are witnessing. Zehner also demonstrates that clean energy solutions are inadequate to address our current needs—not to mention future ones. He targets his message to those who create and adopt neoliberal policy, alongside mainstream environmentalists. In essence, Zehner claims that it is "entirely unreasonable for environmentalists to become the spokespeople for the next generation of ecological disaster machines, such as solar cells, ethanol, and battery-powered vehicles."[67]

Zehner knows that his argument is not easily accepted in a world enchanted by sustainability. Environmentalists often feel betrayed when scholars attack the very few solutions that have captured people's imaginations, as they help create environmental awareness. There are just too many people who benefit from designing and implementing these new technologies. "During times of energy distress, we Americans tend to gravitate toward technological interventions," says Zehner. According to him, Americans should, instead, seriously address "the underlying conditions from which our energy crises arise."[68]

Unsurprisingly, Zehner's book sparked an emotional debate when it was released in 2012.[69] Many in the sustainable development industry accused him of using false figures to show that solar panels and wind power were polluting. Defenders of electric cars alleged that Zehner applied biased data to argue that batteries are dangerous for the environment. But the attacks on *Green Illusions* only reinforced Zehner's argument: we have become overly fixated on the details of technology without acknowledging the real problems surrounding overconsumption, social injustices, and poor policy. In his response to a journalist, Zehner explained, "There's no such thing as clean energy, but there is such a thing as *less* energy."[70]

Another open critic of this enthusiasm for Northern technology to resolve the problems of the South is Chandran Nair, the author I introduced in chapter 5. In *The Sustainable State*, he laments that "we talk about how to make our increasing production and consumption more resource efficient, when we should actually talk about how a global population of 10 billion by 2050 must produce and consume less."[71]

Some commentators position recent attacks on sustainable technologies—those that focus people's attention on more structural changes in behaviors—as a fight against a straw man.[72] They remark that, since the 1960s, there have been scores of environmentalists battling against overconsumption, social injustice, and human rights violations among the most vulnerable. For these commentators, all battles, from modest reductions in kilowatts to radical changes in ways of living, can, and must, be fought at the same time.

Naturally, this is only reasonable to a certain extent. After twenty-six years of working with engineers, urban planners, and architects, I have found that the vast majority of sustainable solutions in construction and city planning have not really challenged our ways of living. Very few have addressed social injustices or reiterated the needs and expectations of the most vulnerable. Most policies under the sustainability agenda have maintained current practices while merely encouraging us to use green gadgets that increase energy consumption and so rarely produce the expected long-term results. When it comes to making decisions after disasters, the findings by Zehner and Nair remain pertinent.

We must consider the disproportionate number of resources devoted to finding innovative, postdisaster solutions, against those used to prevent them. Minimizing people's vulnerabilities through measures that favor social justice requires less money than rebuilding after disasters. Robert Muir-Wood, a professor at University College London and author of the book, *The Cure for Catastrophe*, compares U.S. spending over the past two decades for both disaster prevention and reaction. "Each dollar of extra preparedness spending" he finds, "reduced disaster impacts by an average of 7$ over a single four-year election cycle, and disaster costs overall by an average of 15$." But for politicians, disasters are ideal opportunities to boost their image, he says. "Money spent in preparedness wins no votes."[73] Alternatively, public announcements of massive investment following destruction are used by inventors and companies to make money fast and enlarge their businesses. The public-private partnership at the core of sustainable development seems to work well for the most privileged.

THE EFFICIENCY PARADOX AND THE LOW BAR SET

Some economists and writers have affirmed that increases in efficiency in energy and manufacturing eventually lead to more consumption and waste. One of those writers is David Owen, author of *The Conundrum: How Scientific Innovation, Increased Efficiency, and Good Intentions Can Make Our*

Energy and Climate Problems Worse. In *The Conundrum,* Owen explains that "since the mid-1970s—the period during which refrigerators became bigger, yet cheaper and more efficient—per capita food waste in the United States has increased by half," adding that "we now throw away 40 percent of all the edible food we produce." He paints a similar portrait of electric cars and asserts that, "Global Environmental Enemy no.1 is the automobile, no matter what it runs on." His message for urban planners who design exclusive lanes in highways for electric cars is "any so-called green scheme that makes you happier to be a driver is both delusional and counterproductive." After all, the problem with mobility has been interpreted by the sustainability industry as a ratio of miles per gallon, when the real problem is miles, period.[74]

The benefits obtained by using energy-efficient materials are significantly (and sometimes completely) overridden by increases in consumption. For Owen, this is our critical contemporary "conundrum."

Unlike *Drawdown*, I don't have a panoramic or encyclopedic view of all industries. You may have noticed by now that I know rather little about refrigerators, food packaging, and many other sectors that are responsible for carbon emissions, and—with two children as tangible proof—I'm only slightly better at family planning. But I know that the skepticism surrounding technology in contexts of global warming fits the reality of the city-making and reconstruction industries I am part of. In these sectors, we rely on technology to fortify the status quo—a deceiving approach to prevent unnatural disasters.

The exaggerated emphasis on technology in architecture and urban design already distracts us from basic low-tech principles that are more effective in reducing carbon emissions. In chapters 3 and 4, I describe one of the clearest examples of virtuous damage in architecture that increases reliance on new, sophisticated machines for making buildings work. Jean-Jacques Terrin, a French professor, architect, and planner, calls this "the kilowatt sprint." He says that architects are "more focused on reducing a fraction of a kilowatt per square meter than the overall behaviour of consumerism."[75] Another French architect and environmentalist once told me, "We tend to forget that the only truly sustainable building is the one that is *not* built." An example of pseudo-virtuous acts that I also addressed before is the increasingly common adoption of green certifications. Any competent professional in the industry knows that a building in a remote location, surrounded by highways and an asphalt parking lot, has few environmental merits. And yet, many of these buildings still obtain green certifications in North America. In 2018, a shopping mall located next to a highway in the periphery of the Mexican city of Leon obtained a LEED certification[76]—and all of a sudden, purchases at Zara and Forever 21 began to look greener.

New modes of communication are also playing a role in our failure to engage in serious debates about the sacrifices that we need to make. Our social media platforms are now full of short video clips displaying technological gadgets that can apparently solve the world's problems: "bio-curtains" in buildings' facades that capture CO_2, structures that produce water from air humidity in Africa, machines that remove plastic from the ocean, equipment that cleans rivers in Peru, and bio-materials for disposable construction components in India, among others. These clips wouldn't be so bad if they'd adequately expose the innovations' true (experimental) value. But in most cases, they are presented as fully tested and validated ideas—ready to be used, much like the accessories that Mexicans can now buy in their new LEED-certified mall. The videos try to convince us that we don't need to change much to solve global warming or poverty; that we simply need to wait for developers to start buying these commodified solutions.

When I was invited to give a talk at a sustainable event called Ecosphere in 2018, there were dozens of booths around the conference room that exhibited sustainable innovations. Most of them were physical manifestations of the innovation clips found on social media. It was difficult for me to count the number of "likes" these booths had, but judging from the crowds of hobbyists and enthusiasts surrounding them, I surmised that they were very popular. We can, and must, raise the bar higher.

INNOVATION IN THE CITY—ANOTHER DEBATE AT 30,000 FEET FROM THE GROUND

In chapter 2, I revealed that debates on the effectiveness of aid are often disconnected from realities at ground level. Controversies encircling the role of technology in environmental contexts typically face the same problem. Predictions around the potential benefits of innovation in the city are usually dependent on broad generalizations and untested assumptions often detached from evidence obtained on the ground. One clear example of this is the relationship perceived between land and solar energy.

In 2013, the U.S. National Renewable Energy Laboratory (NREL) published a study about land-use requirements for solar power plants. The study detailed the size and characteristics of land needed for solar panel farms, without offering any distinct recommendations. But later interpretations of the data calculated that "a solar power plant that provides all of the electricity for 1,000 homes would require 32 acres of land."[77] Some commentators concluded that switching to solar power did not make any environmental sense. The strategy would put pressure on the value of urban parcels, reduce agricultural space, and decrease urban densities. All of this would impact housing affordability, destroy natural species, diminish food production, and compromise wild land.

Nonetheless, defenders of solar energy were not convinced by these arguments. Many pointed to an earlier study published by the NREL indicating that photovoltaic energy "would offer a landscape almost indistinguishable from the landscape we know today."[78] The early report argues that urbanized areas in the United States cover about 140 million acres of land. "We could supply every kilowatt-hour of our nation's current electricity requirements simply by applying photovoltaic (PV) to 7 percent of this area," determined the report. The authors found that by relying on panels exclusively installed on rooftops and vacant areas of the city, developers "wouldn't have to appropriate a single acre of new land to make PV [their] primary energy source!"[79] (It is perhaps worth noting that solar farms and rooftop solar initiatives occupy the eighth and tenth positions in *Drawdown*'s list).

Both sides of this debate appear disturbingly hyperbolic compared with our actual knowledge of contemporary architecture and cities. A more nuanced argument emerges from an analysis at ground level. Take, for example, when the city of Montreal built a net zero facility for a public park called Gouin in 2017. The two-story building was expected to produce only as much energy as it consumes in a single year (hence the term "net zero building" in the industry). The initial drawings depicted a set of photovoltaic panels on the roof. But the designers soon realized that it was difficult to reconcile the location and orientation of the building, the design of the roof, and the aesthetics of it all, with the size and alignment required for the solar panels. As such, they opted to build a separate structure, with the right size and orientation to install these panels. While it is true that no agricultural land was scarified for this endeavor, the project in Gouin involved building in a park. It would be a different story if the structure was built in a denser location. It is also true that the structure, which was built in addition to the main building (and surely consumed urban space and energy to be assembled), doesn't have any significant purpose aside from supporting the solar panels. In the case of Gouin building, solar energy is neither a magic solution nor a dangerous one. Like many other sustainable gadgets, the effectiveness of solar panels depends on the context in which they are applied.

I believe that the majority of professionals in construction and real estate adopt green gadgets in buildings and infrastructure with honest intentions. But incorporating technology without considering its specific characteristics or endorsing untested practices is not ideal for overcoming our environmental anxiety.

RISK IS NOT EQUAL FOR ALL—EVEN IF YOU HAVE THE BEST APP

Throughout this book I have shown examples of gentrification, displacement, and other situations where rapid urban change has occurred and resulted in unexpected consequences for the most vulnerable. But in case you are still not fully convinced of how difficult it is to anticipate the detriment of certain forms of innovation on poor, marginalized, and excluded groups, consider one more example: the emergence and widespread adoption of apps for space sharing.

A few years ago, many of us celebrated the arrival of Airbnb and other apps that help us share unused space. This innovation was initially seen as a sustainable way of optimizing existing infrastructure and buildings. It was embraced by many as a true form of solidarity economy. As the years went by, however, a different portrait of Airbnb emerged.

A study by David Wachsmuth, a professor at McGill University, recently determined that Airbnb has contributed to the loss of thirty-one thousand homes from Canada's rental market.[80] Increases (albeit small) in Arbnb offerings are already raising rents in Toronto, Vancouver, Montreal, and other Canadian cities.[81] The most affected are, of course, the poor. Social groups argue that Airbnb is distorting prices on the market.

Wachsmuth's findings match those in several other cities.[82] Today, cities like Amsterdam, Barcelona, Berlin, London, and New York are imposing restrictions to reduce the housing commodification amplified by Airbnb.[83] Innovations like Airbnb and Uber—once viewed as sustainable leaders in the sharing economy—quickly mutated into multinational enterprises with the power to challenge regulations and laws in many countries. The companies behind these innovations have enjoyed several benefits. A major one is avoiding the taxes being charged to their traditional competitors. In the process, they have become multibillion dollar corporations with economic power and political influence. A network of big companies certainly benefits from online platforms.

A study conducted by CBC in Canada found that 44 percent of Airbnb hosts are, in reality, companies with more than one unit listed. About 22 percent have at least five listings, so a great fraction of those listed on Airbnb are effectively hotels "operating out of what used to be, or what otherwise could be, people's homes," said Wachsmuth in an interview.[84] Unsurprisingly, labor unions, traditional associations, and other social groups in Canada are concerned about the deterioration of working conditions and unjust business practices brought about by these platforms. The most vulnerable Canadians are paying the price.

Most economists and defenders of innovation tend to argue that criticism of radical, technological change is nothing more than a familiar resistance to change. In most contemporary societies, resistance to change fueled by technology is increasingly treated as an error to be corrected in the same way that a sick person must be healed. In 2016, Harvard professor Calestous Juma, wrote a celebrated book called *Innovation and Its Enemies: Why People Resist New Technologies.* Just from the title of this book, you can get a sense of Juma's argument: critics of new technologies are often depicted as enemies of innovation—short-minded conservatives who fail to see the broader picture, the long-term benefits, and the beauty of new solutions. In some passages, Juma deplores how in debates about technology, the legitimate concerns of the most vulnerable—those who consider themselves to be on the losing end of innovation—are dismissed. But he also believes that most debates are "about perceptions of risk, not necessarily about the impact of the risks themselves." For him, these "controversies grow out of distrust in public and private institutions."[85] The issue is that, as in cases presented throughout this book, the concerns of the most vulnerable have repeatedly been dismissed by calling them "simple perceptions" or "blunt distrust."

When it comes to links between disasters and technology, defenders of radical change also claim that criticism of innovation misses three essential points. One is that many past radical changes have *emerged* in the aftermath of major tragedies. The notion is that tragedies and destruction spark our creativity, inviting us to push ideas to new frontiers. The development of several safety measures, medicines, and treatments, for instance, can be tracked down to terrorist attacks, wars, and epidemics. But these defenders of innovation forget that numerous mistakes have also been made in the name of efficiency. It is easy to remember the success stories of innovation, yet the tragic and more problematic ones usually leave invisible traces. Some of these include the thousands of cases of pulmonary cancer caused by breathing in fragments of prefab panels made of asbestos—a ground-breaking innovation in construction in the 1970s.

The second argument is that disasters are often followed by sharp increases in cost, due to speculation regarding the price of materials and labor. Defenders of innovation argue that new solutions create competition, which can help reduce costs and enhance quality. Increases in construction costs after disasters are, indeed, an incentive for companies to enter new markets and, to be sure, there is nothing wrong with competition and companies trying to make money. But three key problems still remain. One is that new solutions rarely respond to people's needs, and so they fail to address their desires and expectations. Second, new solutions are regularly backed by politicians and authority figures who benefit directly from support that is not proportionately given to more vernacular or indigenous solutions. Finally, the search for postdisaster innovation persuades companies that want to *test* ideas to use disaster victims as guinea pigs. Enthusiasm for innovation sometimes helps those corporations that have been unable to successfully market their products to capitalize on the disruption caused by disasters, and thus, to sell them with less resistance.

Defenders of radical change frequently argue that product innovation is often accompanied by the creation of institutions that bring long-term value to society. "Technology and institutions are as inseparable as institutions and technology," writes Juma. He contends that the long-term benefits of institutions offset the short-term collateral effects of new products.[86] But again, we must remember that negative effects are not distributed evenly—less so in times of disasters. Do we really want the sacrifices and costs to fall on the most vulnerable today for the mere possibility of benefits in an indeterminate future? Wouldn't it be fair to have the most privileged assume these risks now so that there are far less vulnerable people tomorrow?

PLAYING WITH FIRE

We continue to underestimate the risks of technological innovation. We forget that the atmosphere and ecosystems are incredibly complex systems. Never before has disaster creation become more literal than with recent innovations for geoengineering aimed at halting climate change. Some of the most absurd solutions include artificially altering the atmosphere, as though playing God. Proponents argue—some with a straight face—that a fleet of airplanes could start spraying a fine mist of sulphates, calcite, or nanoparticles into the atmosphere, which would "reflect back just enough sunlight to prevent dangerous warming," writes Steven Pinker. In *Enlightenment Now*, Pinker asserts that this would replicate a major disaster, like "a volcanic eruption such as that of Mount Pinatubo in the Philippines in 1991."[87] Needless to say, we can't actually know the consequences of these radical innovations on plants, animals, and humans—not now or for decades to come. And yet, a growing number of people are taking them seriously.[88]

Innovation enthusiasts also tend to forget that cities are highly complex systems. Due to this complexity, it is difficult to anticipate the full impact of innovations on different social groups. For the wealthy and powerful, innovations in housing, transportation, and other services are perhaps a risk worth taking. But for most vulnerable people, the effects of innovation can be catastrophic.

Most innovators (and fans of Steve Jobs) like to say that "the best way to predict the future is to change it." But taking this risk on behalf of those who lack a safety net is irresponsible. In chapter 2, I identify the role of consent as the main difference between design in a city and the design of objects and accessories. This same principle applies to innovation in a world of unnatural disasters. Several advancements made in housing, transportation, public space, and infrastructure affect both the private and public lives of the most vulnerable. They also impact vulnerable people who might not have given their consent. For the poor, marginalized, and excluded—whose voices are often less heard—the stakes are simply higher.

This does not mean that we must stop trying to solve these pressing issues. The argument of this book stresses the exact opposite. What it does mean is that the precautionary principle should guide any risky decisions that will particularly affect the most vulnerable. Overconfidence in high technology compromises this principle and underestimates the risks involved. In chapters 7 and 8, we will explore solutions that do not require sophisticated machines, high-tech materials, or software—only the ingenuity of traditional materials and vernacular solutions.

Since sustainable development became popular, our intentions to protect nature have come with an underlying message: let's find something new to sell, while we try (or pretend) to save the planet. This has distracted us from a serious debate about lifestyle, our modes of living. We have to come to terms with the fact that radical change is fuelled by savage capitalism. Capitalism depends on what economists and engineers call "creative destruction"; that is, the constant replacement of technologies as new ideas emerge—a process that feeds on consumption.[89] That is why defenders celebrate—and critics deplore—technological innovation as a convenient way to maintain economic growth.

By 2050, the world will have 9.8 billion people. That is 2.3 billion more people than there are today.[90] It is projected that 6.7 billion will live in cities. Some of the most rapid (and unimaginable) changes will occur in Africa and Asia, where urbanization is still on the rise. The demands for energy, food, housing, materials, water, and space will be enormous. So will be the amounts of pollution. In order to avoid disasters, we need new solutions across a wide variety of sectors. The search for new solutions is a laudable objective, but using disasters and disaster-affected populations as laboratories to test new technologies is not.

In terms of climate change, serious journalists, scholars, politicians, and environmentalists alike will tell you that "there is no magic bullet." Technological innovation will surely be pivotal for dealing with environmental and social problems. What we must realize, though, is that technology won't solve them unless we consume less and take care of our most vulnerable, give more to them, and build neighborhoods and infrastructure in ways that correspond to their traditions, rituals, memories, and needs. We must celebrate that some people will be successful through innovation. Yet we cannot forget that there are also plenty on the losing end. In order to help protect them, we must pay more attention to what they have to say.

Allora, hanno cambiato
il logos del loro viaggio
Che non è più un attracco
Né
Una locuzione ma una
Ricostruzione.

7 | Decision-Making

"We Want to Be Able to Make Our Own Decisions"

It pointed out, doctrinally, that the Lottery is an interpolation of chance into the order of the universe, and observed that to accept errors is to strengthen chance, not contravene it.

—Jorge Luis Borges, "The Lottery in Babylon"

NOT IN YOUR EYES IN COMUNA 13

"Forget what you have seen on *Narcos*,[1] and similar shows on Colombian drug cartels," says Alejandra, our guide during a visit to Comuna 13, one of the slums in Medellín. Colombia.

Alejandra grew up in Comuna 13. Just a few years ago, this was one of the most notorious neighborhoods in Colombia. She knows that most foreigners identify Medellín as the capital of drug trafficking, and she feels both offended and ashamed by this reputation. Alejandra is about twenty-five years old, intelligent, and—at first glance—self-confident. She definitely doesn't want to be associated with the sexually objectified women typically portrayed on *Narcos* and most films about mafias. She wants to exemplify a new generation of young, educated, and empowered Colombian women. Alejandra wants us to think differently about her country, the low-income *comunas* and *paisas* (as people from Medellín and the region are often called).

As a kid, Alejandra witnessed street violence of the worst kind. Until very recently, she couldn't walk freely through the slum. There were too many invisible frontiers—limits imposed by local cartels to control the territory. "You were murdered or raped if you entered a neighborhood or sector that was controlled by a certain mafia," she told me. Now she has a part-time job with the municipality, leading three to four groups of tourists per day through Comuna 13. She is proud to describe how the place has changed over the past twelve years. Alejandra guides tourists through the open-air electric escalators that were built to access the higher points of the comuna. She delivers valuable information about how difficult it was to build sidewalks and infrastructure due to the sloped geography of the slum. Alejandra has a story for each of the shops and hip cafés that have recently opened in Comuna 13. She also explains the hidden messages behind many of the graffiti murals that decorate the new retention walls and homes as we ascend the comuna to enjoy the beautiful views of the slums.

But many in our group were more interested in *her* story than in the explanations for new sidewalks, metro lines, and escalators in Medellín. How did she get this job? Where did she study?

A few years ago, the city and some charities ran an education program for young people in Comuna 13. Alejandra enrolled as a way to escape boredom, and she took several courses in communication, history, and public speech. With that training, she got her first job as a tour guide in the comuna. She knows the technical jargon of urban design and provides tourists with a rich history of projects in the slums, which made our visit all the more enjoyable.

At first glance, Alejandra looks self-assured in addressing our group. But she has something to confess:

> "I never look at strangers in their eyes. It is something I still need to work on. I can talk in front of a dozen strangers, but never look one in their eyes."

In speaking with Alejandra, you realize that it is easier to transform sidewalks in a slum than it is to turn fear into trust. But there has been both a physical and a human transformation in Comuna 13. Not many years ago, young women from the slums had little hope of making an honest living. Marrying a gang member with money was one of the best means to escape poverty. The most likely path for male teenagers, meanwhile, was ascending the hierarchical ladder of local cartels—a career that demands full-time violence and has two possible ends: death or jail.

In the 1990s, Medellín's low-income comunas were largely disconnected from the urban fabric of society. They had become incubators of organized crime. Even the police could not easily access the slum's labyrinth of houses and narrow dirt roads, where notorious mafiosos and gang members hid. Disasters in the poor comunas had become just as common as the violence. In 1987, when the Medellín drug cartel was one of the most powerful in the world, a landslide buried 562 people and about 270 homes in Comuna 8, another notorious slum at the time.[2] In 2007, a landslide killed eight people in Comuna 13, including three children.[3] In 2001, 2007, and 2010, massive fires destroyed hundreds of homes in the area.[4] The weak constructions of the slums built on the mountains made them fragile and more susceptible to landslides in the rainy season and fires in the dry season. With global warming, disasters in Colombian slums have also become more frequent and intense. In the rainy season, water surges have become a major problem, particularly for those living in weak structures. In the dry season, fires are a common threat for both slum dwellers and the forests around the low-income comunas.

"Slum dwellers didn't think of Medellín as their own city," said Margarita Inés Restrepo, an expert in disaster-risk reaction and the former leader of a local NGO called Corporación Antioquia Presente. This was a manifestation of the exclusion felt by slum dwellers for many years. "Slum dwellers often talked about *going* to Medellín, even when they lived in the immediate periphery of the city."

Today, most slums in Medellín, including Comunas 13 and 8, are safer. Disasters are not as prevalent and less deadly. To be sure, crime and violence are still common but have largely receded from their peak in the 1990s. Whereas, in 1991, there were 6,349 homicides in Medellín, that figure was reduced to 1,044 in 2008—even as the population increased from 1.6 to 2.3 million during that period.[5]

There has been an unprecedented urban transformation in the slums, largely fuelled by new public transit solutions. Medellín's first cable car system, which opened in 2004, is one of them. This gondola lift runs over about 15 kilometers, connecting the highest points of the comunas with the metro system.[6] There are also state-of-the-art schools and public libraries, including the Biblioteca España, designed by a prominent Colombian architect and partially funded by the Spanish Cooperation Agency. Several low-income settlements, including comunas 8 and 13, now have paved roads, parks, sports facilities, and sidewalks. Public space in the low-income comunas features modern urban furniture, new lighting, and local street art.

Because of this, researchers and visitors from all over the world see Medellín as an example to follow. Scores of tourists now visit the area to enjoy and understand its impressive transformation from capital of drug traffic to the capital of urban solutions. In 2012, the *Wall Street Journal* named Medellín "Innovative City of the Year."[7] Other cities in Colombia and Latin America are trying to emulate its success, resulting in what is called the "Medellín effect." Cali and Bogotá have built cable car connections between the slums and public transit lines. Bogotá has also built a network of libraries, many of which are found in slums. The capital city wants to copy Medellín's public transit success, which has prompted the construction of cable cars in Ecatepec de Morelos (near Mexico City), Caracas, La Paz, and Rio de Janeiro. There are plans to build similar systems in other cities across the Global South.[8]

Medellín has become a transit zone for almost all the international urban consultants who have popularized sustainability and resilience discourse worldwide. Many consider the city's social urbanism an urban miracle. Yet, to be fair, not everything in Medellín has worked as expected. Crime, domestic violence, sexual abuse, and drug trafficking are still major problems in the low-income comunas. Minor landslides and fires in the area have also been reported in recent years. The development of a green belt, aimed at reducing risk and preserving green space in the mountains, has been largely criticized for displacing low-income residents. Scholars and social leaders have declared the green belt a beautification project for the rich,[9] while civil society groups have contested the relocation proposed by developers of the initiative. Many resent that the rich are always given the privilege of occupying more and more suburban land, but when poor neighborhoods expand, authorities claim that they are causing an environmental crisis. Besides, when the rich occupy sloped land, authorities pay for additional infrastructure, such as retention walls, to ensure their safety, but when the poor do it, authorities refuse to connect services and instead label the locations as "disaster-prone."[10]

Other problems persist. Inequality, for instance, is still a major challenge in Medellín. An enormous gap keeps widening between the living standards of the rich, who live in posh neighborhoods like El Poblado, and the residents of the slums. Administrative scandals and technical problems have also surrounded certain initiatives. A few years after its official opening, the Biblioteca España was closed due to structural and construction deficiencies.[11] Air pollution is another threat in Medellín; it ranks among the highest in Colombian cities.

Two years after we met Alejandra in Comuna 13, I received sad news. She had been arrested and accused of killing a police officer. Authorities suspect that Alejandra was both the romantic and criminal partner of a local murderer. She is now in jail awaiting trial.

Despite the corrosive impact of the war on drugs and the fifty-year civil war in Colombia, Medellín has done much more than other Colombian and Latin American cities to reduce the vulnerabilities of slum dwellers. As a frequent critic of rapid change imposed in cities, the transformation of Medellín has always puzzled me. There are so many innovations in Medellín's urban transformation, ranging from novelties in mobility, urban furniture, architecture, and communication. As the public escalators suggest, technology has been used in creative ways, which begs the question: Why has innovation succeeded in Medellín, but failed in the cases we examined in previous chapters?

Juan Miguel Pulgarín, a social worker who participated in the transformation of both Comuna 13 and Comuna 8, seems to have part of the answer.

INSTITUTIONAL ZOMBIES AND CHANGE IN THE SLUMS

Juan Miguel has a master's degree in urban studies and is a consultant for the Empresa de Desarrollo Urbano de Medellín (EDU), a public company that procures and builds urban projects in the city. "You don't obtain long-term positive change through physical infrastructure alone," Pulgarín told me in an interview in 2019. He was in charge of several urban development projects during the administration of Sergio Fajardo, a popular mayor of Medellín from 2004 to 2007 who then became governor of the Antioquia region (where Medellín is located) from 2012 to 2016. "Many foreigners come and see the parks, cable cars, escalators, and tramway in the comunas and believe that they are solutions to the problems we face," Pulgarín explained to me. "But these are *not* solutions, these are the *consequences* of structural programs that Fajardo and other mayors put in place years before."

For Pulgarín, long-term positive change only comes with education, social programs, efficient policing, a shared vision, and collaborative work with citizens and social leaders. Only then is it possible to pave roads, reduce risk, and build new infrastructure. Pulgarín deplores that people believe it is the other way round. He regrets that city planners and consultants do not recognize the enormous effort that underlies urban transformation; effort that builds slowly and demands both institutional engagement and energy. Most of the actual solutions, such as education, information, and public awareness, are not easily perceived by tourists but are crucial to the visible results seen today.

Pulgarín believes that the new roads, sidewalks, parks, sport facilities, and libraries built in the Medellín's low-income comunas are just the tip of the iceberg in the overarching program to integrate slums in the city. "It is not about roads and buildings," he goes on. "It is about being able to maintain a vision of social change over several years."

But why is it possible to maintain this lasting vision in some cases but not in others? The answer lies in strong institutions.

In the second half of the twentieth century, Medellín focused on promoting local industrial development instead of imports. The city became a hub for powerful factories in the Latin American textile sector. The city also consolidated a series of public-owned utility firms into one company called

Empresas Públicas de Medellín (EPM), which provides electricity, gas, water, sanitation, and telecommunications services. Medellín made significant strides to protect the local economy. When traditional industries, such as textiles, became less productive (in part, due to cheaper imports from Asia), Medellín reinvented itself as an axis of new technologies.[12]

The capacity to integrate good practices and utilities under one company made EPM more effective and well-respected. EPM became an anomaly. In the 1990s, most public-owned companies in Colombia were ill-staffed and virtually bankrupt. A high-ranking officer once said that public utility companies in the country were nothing more than "institutional zombies." That is because the implementation of neoliberal policy aimed at reducing investments in state-owned companies and government agencies. But as the policies of the Washington consensus were implemented in Bogotá, Cali, and other Colombian cities, Medellín resisted. Of course, there were several attempts to privatize the EPM, but they failed under the pressure of labor unions and businesspeople.[13] According to Julio Dávila and Peter Brand, two experts in urban planning, the success of the public transport system in Medellín is "the result of a culture of political leadership at the regional level, and strong public institutions that resisted the privatization movement of recent decades."[14]

Part of EPM's success in defying privatization derives from what Colombians call *regionalismo*, or the tendency to provide significant value to regional traditions, kinship, and businesses. Paisas are known for being frequent practitioners of this type of regionalism; a trait that facilitates the emergence and maintenance of strong companies and public institutions. Unions and general citizens alike are proud of the EPM and protect it as a valuable facet of Medellín's local heritage.[15]

At a time when many municipalities in Latin America privatized their public utilities, Medellín converted EPM into a multinational corporation. EPM is now present in Colombia, Panama, Chile, El Salvador, and Guatemala, among other countries. In 2016, up to 35 percent of its revenues came from international operations.[16] Mainstream economists often struggle to explain the functionality of the EPM. It is run as a commercial enterprise and managed by the municipality of Medellín, while preserving close ties to the mayoral office. It is an autonomous economic entity, yet Medellín's mayor serves as president of the company and appoints its board members.[17] The EPM also keeps decision making independent and has largely avoided political manipulation. The company was recently valued at more than 10 billion dollars.[18]

EPM has invested in most of the urban revitalization programs associated with the transformation of Medellín. In March of 2012, the company launched a US$50 million private equity fund to leverage business developments in innovation, science, and technology.[19] To be sure, EPM has also invested in major infrastructure projects, some of which have been controversial. For instance, the company has been criticized for its involvement in the construction of Hidrotuango, a massive dam in the Cauca River, one of the largest rivers in the country. Hidrotuango broke in 2018, causing a major environmental disaster in the region. But the popularity of EPM remains high. In 2019, an editorial in *El Colombiano*, a popular paper, called it a "crown's jewel."[20] It has been a key player in upgrading programs within the low-income comunas.

Another significant ingredient in the transformation of Medellín has been the city's capacity to attract investment and aid. In 2001, the city created a cooperation agency called the Agencia de Cooperación e Inversión de Medellín y el Área Metropolitana, or ACI.[21] Ten years later, an ambitious international cooperation policy was approved by the city. Since then, ACI has established partnerships with dozens of international institutions and associations. Between 2004 and 2017, they amassed more than US$90.6 million in international cooperation—with about 20 percent of those resources going to urban initiatives.[22]

Another strong, public-owned company is the Empresa de Transporte Masivo del Valle de Aburrá, which operates Medellín's metro system. This is a growing company that works with investments from the city and regional government.[23] It is also one of the strongest brands in the country. The Medellín metro transports eight hundred thousand people every day, 92 percent of who are low- and medium-income citizens.[24] "Before building the Metro, Medellín conducted an education program to create awareness among the public," Margarita Inés Restrepo explained to me in an interview in 2019. "When the transportation system was opened, paisas already wanted the system and were ready to protect it."

During the Fajardo administration, the city also assured that infrastructure development would be accompanied by participatory budgets. Between 5 and 7 percent of the municipal budget was controlled by residents.[25] With those resources, civil society organizations and residents of the low-income comunas were able to assume a leading role in managing and planning local initiatives, often involving public space and infrastructure. For many years, Margarita Inés Restrepo was in charge of environmental initiatives in Medellín. "Participatory budgets created the space for social debate," she told me in the interview. "There were intense discussions and negotiations within the comunas to decide how to invest resources. This gave slum dwellers an opportunity to engage in peaceful dialogue and a common vision."

While there have certainly been international consultants involved in Medellín (the Rockefeller Foundation, UN Habitat, and ICLEI, to name a few), local NGOs—independent from international agencies—such as the Corporación Antioquia Presente, have been key in the urban transformation process. When disasters in the slums were all too frequent, Antioquia Presente supported the social change required for physical reconstruction. The model, called Desarrollo Social Integral, or Integrated Social Development, was based on education, training, and the development of trust between social groups and institutions. NGOs played a fundamental role in supporting what became known as Proyectos Urbanos Integrales (or Integrated Urban Projects). Before the 1990s, charities working in Medellín were predominantly funded by international agencies and private donors, but by 2008, 70 percent of their funds were provided by the state.[26] Margarita Inés Restrepo explained to me: "Charities helped to provide social support needed for the physical transformation of the city."

Investment in culture was equally critical. In 2003 only 0.6 percent of the municipal budget was devoted to cultural initiatives, but by 2008, 5 percent was dedicated to various forms of cultural expression and art. Music, dance, and street art became a kind of social resistance against violence and neglect. Comuna 13, for example, is now trademarked. You can buy t-shirts, caps, and other accessories with its logo. This is proof that Alejandra and many other paisas from the slums want tourists to spend money in their comunas. But it is also one of the most recent efforts to make tourists forget about the stereotypes promoted by *Narcos* and in the media.

To be sure, Medellín's transformation still faces substantial challenges. As I write these lines, the city is struggling to cope with the COVID-19 pandemic and there is also a new corruption scandal surrounding EPM. But the city has an institutional infrastructure capable of dealing with several challenges in the future. As for Pulgarín, he is still working on urban projects in low-income communities, whereas Clara and Margarita Inés Restrepo are now retired. In 2017, Fajardo ran for the Colombian presidency and finished in third place. He is still a powerful politician.

THREE LESSONS FROM MEDELLÍN

The transformation of Medellín teaches us three important lessons that are crucial in times of unnatural disasters: First, the measures taken in the city contradict the neoliberal teachings that Washington, London, the World Bank, and the IFM have tried to implement in developing countries for decades. Neoliberal measures have mostly created disaster risk in Latin America, whereas the protection of public institutions in Medellín generated conditions that have actually *reduced* risks among the most vulnerable.

The majority of decisions made in Medellín wouldn't have succeeded with weaker public organizations. Strong institutions are necessary to make difficult choices in investment, environmental protection, and urban planning. They have also been key in managing pressure from economic groups and guarding checks and balances to avoid corruption. By using these strong institutions as investors, mayors such as Fajardo have been able to focus on correcting historic mistakes that initially led to the marginalization and segregation of social groups. Surely, participation was a vital component of the Fajardo administration, but this was not done through the superficial involvement of token audiences in decisions that had already been made. Participation during the Fajardo administration was a space for delegating power and allocating budgets to civil society groups so they could make decisions according to their own objectives and desires.

Second, the Medellín case teaches us a lesson about innovation in the city and in response to environmental challenges. Most decision makers believe that innovation is the vehicle that produces change. While a few sustainable developers and consultants recognize that innovation carries short-term negative effects, they quickly add that rewards are obtained in the long term. As innovations continue to create or consolidate social institutions, they concomitantly enhance social benefits. However, the Medellín case shows that the opposite is true. Innovation is not the *cause* of positive, long-term structural urban transformation—as most urban consultants would like us to believe. If anything, it is the *result* of engaged and sustained work over long periods of time. And this sustained work is only possible when institutions are strong. Innovation emerges as a response to urban and environmental problems when there is an institutional environment capable of mobilizing resources, connecting stakeholders, developing a shared vision, and establishing and maintaining implementation mechanisms.

Finally, this case confirms that it is better to reinforce existing settlements and slums than to replace them with new housing developments. Governments and charities often seek to replace informal settlements with residential developments or green areas. A common argument is that this helps prevent disasters, increases security for slum dwellers, and improves their living conditions. But as we have seen in earlier chapters, this strategy increases the vulnerabilities of low-income residents who find themselves displaced and living in units that are ill-adapted to their needs and expectations. In the Medellín low-income comunas, the main strategy has been different.[27] Efforts have been made to integrate slums into the city, rather than bulldozing or replacing them with new developments. Proper integration of poor neighborhoods thus enables access to opportunities in the city, while respecting attachments to local social networks and their neighborhoods.

As we shall see, the benefits of improving infrastructure instead of merely replacing it with new solutions applies in rural contexts too.

COFFEE-QUAKE

Like most of my fellow researchers and humanitarian practitioners, I got into the disaster field convinced that I could develop a technical solution after a period of mass destruction. The event that triggered this idea was an earthquake in rural Colombia in 1999. The disaster killed almost a thousand people, mostly peasants and citizens in small towns. It also destroyed thousands of homes and had a strong impact on the national economy.

In 2000, I traveled to La Tebaida, Circacia, Montenegro, and other towns affected by the earthquake. These towns are rich in vernacular architecture and sit on a beautiful landscape of coffee plantations, rivers, and green mountains. The region is known as *eje cafetero*, as its main economic activity is the production of coffee. The eje cafetero, or coffee axis, stretches from the Medellín region of Antionquia to five other departments, or Colombian regional jurisdictions. I met several rural residents who had been affected, in one way or another, by the disaster. Some lost family members, others saw their small businesses shattered, or had their homes completely or partially damaged.

When I met Mr. José Bermudez, he was proud to show me his home. A few years before, his children had left for Medellín to work and study. He was living alone with his wife, and together they ran a business harvesting and roasting coffee beans. The roof in the kitchen of his home had collapsed during the disaster, so he had rebuilt a new room thanks to a grant provided by a reconstruction program called Forecafé. José was pleased with his new kitchen counter, storage area, and window next to the kitchen sink (see figure 7.1). He also improved the driveway to his farm and rebuilt the septic tank. He was back in business and hoped to sell his produce in the coming months.

Much like postdisaster Haiti twenty years later, a housing exhibition was built near La Tebaida, where José and his wife live. The best construction companies in Colombia were invited to exhibit their most innovative housing solutions. But José was not interested in buying a brand new home. He liked the house that his family had built with vernacular technologies, decades ago, and he focused on repairing the rooms that were affected in order to preserve the existing structure as much as possible. José eventually visited the housing exhibition, but not to buy a house. "I went to the exhibition to see if I could find good ideas for rebuilding my own home," he told me.

A few days later, in Circacia, I met Ms. Constanza Sanchez, a mother of three, who had similarly been affected by the disaster. Constanza had also received financial support from Forecafé, but she faced a different problem:

FIGURE 7.1 (*Left*) José Bermudez, a coffee grower affected by the 1999 earthquake in Colombia, and (*right*) his newly renovated kitchen.

(Republished with permission of Taylor & Francis from Gonzalo Lizarralde, *The Invisible Houses: Rethinking and Designing Low-Cost Housing in Developing Countries* [New York: Routledge, 2014], permission conveyed through Copyright Clearance Center.)

While her house had not been affected by the earthquake, the *beneficiadero* in her farm had. Beneficiaderos are of prime importance to coffee growers like Constanza. They are two-story structures where coffee beans are cleaned, dried, packed, and stored. Constanza rebuilt the beneficiadero in her farm, and much like José, waited impatiently to sell the years' produce of coffee beans (see figure 7.2). Through this sale, she was expecting to make her annual salary and invest in the following year's harvest.

José and Constanza are part of thousands of Colombians who benefitted from the Forecafé reconstruction program. Through this program, 9,800 rural houses were rebuilt, including almost seven thousand coffee growers' homes. About 4,700 beneficiaderos and other structures for

FIGURE 7.2 Ms. Sanchez's beneficiadero, rebuilt after the 1999 disaster in rural Colombia.

(Republished with permission of Taylor & Francis from Gonzalo Lizarralde, Cassidy Johnson, and Colin Davidson, eds., *Rebuilding After Disasters: From Emergency to Sustainability* [New York: Routledge, 2009], permission conveyed through Copyright Clearance Center)

coffee production were also restored, while close to two thousand families repaired their sewage and water connections. It is estimated that about ten thousand direct and indirect jobs were created during the reconstruction of the eje cafetero.

In 2000, I traveled to Bogotá to meet the leaders of this miraculous reconstruction process. I first met José Fernando Botero in the headquarters of the Federación Nacional de Cafeteros de Colombia, a countrywide Guild or Federation of Coffee Growers. Botero was one of the engineers in charge of coordinating reconstruction activities after the 1999 disaster. His organization had been mandated by the newly-created institution FOREC to lead the reconstruction of rural areas.

FOREC, the Fondo para la Reconstruccion del Eje Cafetero, was an invention of the national government. It arose a few weeks after the earthquake and was in charge of outsourcing and managing reconstruction. FOREC was funded by the World Bank, the Inter-American Development Bank, private donations, and the National Budget—through new taxes that were imposed for recovery. Officials in the central government soon realized that construction activities and resource management could not be handled by a single organization, so they decided to decentralize project operations. They sent out a call for proposals and selected thirty-two NGOs to execute reconstruction projects in specific sectors. By delegating the work to various NGOs, FOREC aimed to decentralize work and resources as well as transfer decision-making power to the organizations that could best handle them "on the ground." It was also an attempt to reduce bureaucracy while creating opportunities for the direct engagement of affected population.

FOREC officials delegated rural reconstruction to the Federation of Coffee Growers. The federation was, naturally, concerned by reductions in coffee production after the disaster and the effects of the quake on the guild's rural members. It was thus committed to helping them and fortifying rapid economic recovery. Besides, the federation was willing to invest its own resources, and it benefited from the credibility and legitimacy of rural peasants. It also had a comprehensive database of coffee growers and a strong understanding of the specific working conditions in rural areas.

There was one problem though: "We had neither the staff nor the knowledge to design, plan, and build thousands of houses and infrastructure projects," explained Botero. The federation then opted for an alternative strategy: to focus on managing the funds and controlling the quality of construction work. The work was to be undertaken by affected households. Instead of providing complete and fully finished houses and infrastructure, the federation relied on what experts now call "owner-driven procurement." In this model, each affected family was eligible for a subsidy of US$4,000 and a US$1,000 government loan. Households were responsible for making their own decisions regarding the use of resources and the construction that they wanted.[28]

Affected families were able to apply for subsidies by proposing one or various individual projects, which could extend beyond their home. Almost any type of investment in infrastructure, roads, or services was eligible for economic support. They simply had to present a photograph of the affected structure or infrastructure and propose a basic design—usually a drawing presented on the back of an envelope. The federation appointed a group of engineers to revise and complete the designs with construction details. Beneficiaries would then either build by themselves, commission formal or informal construction firms, or hire labor alone. They could also use any material or technology that they wanted for their own constructions.

Beneficiaries were placed in the driver's seat. They received subsidies and credits but managed the resources according to their needs and expectations. "The notion of buying a house by catalogue does not exist among most low-income residents and peasants," observed Botero. "They build the family home over many years, sometimes even two or three generations."

"So, why did you build the housing exhibition?" I asked.

"We feared speculation in the price of construction materials and indiscriminate use of wood," he said. Officers in the federation selected innovative solutions based on the quality of the construction system and price.[29] Given that the government offered tax incentives for businesses operating in the disaster-affected area, construction companies offered competitive prices. But the reality was that very few fully finished units were sold. Instead, hundreds of residents like José attended the exhibition to purchase materials and find inspiration to implement the solutions themselves.

Botero and other officers put mechanisms in place to control the quality of outcomes and support residents throughout their decision-making process. A financial institution verified household eligibility to subsidies and credit. After clearance, a team of engineers inspected the individual projects proposed by each beneficiary and verified whether they complied to disaster-resistant and environmental principles, such as avoiding water pollution. After this second clearance, residents were given their first instalment, with which they had to make significant progress before receiving a second and third inspection (about 25 percent of the work had to be completed after each instalment). A process of inspection and approval was repeated as many as four times until the work was done.

In order to support affected families, two construction manuals were published and distributed among peasants. Engineers also provided them with constant information about available resources, construction best practices, and other practical solutions. Furthermore, there was a counselling program led by a group of psychologists offered to the victims. Much like in the Medellín case, all this social support was as important as the cash given to beneficiaries. It not only guaranteed justice in the distribution of benefits, but it also ensured that residents met the technical conditions for reconstruction.

After the reconstruction, housing conditions improved, economic recovery was quick, and there was an overall sense of normalcy for the first time since the disaster. But perhaps the clearest evidence of improvement occurred in 2004. That year, a 5.2 magnitude earthquake hit the region again, but this time, there were no deaths, destruction, or physical damages.[30]

The reconstruction project in the coffee-growing area of Colombia reminds us that disaster victims often prefer to receive cash instead of houses. This tendency has been found in other countries, too. After the 2010 earthquake in Haiti, the American Red Cross distributed cash subsidies to disaster victims. This gave beneficiaries sufficient freedom to find their own solutions and prioritize what was important to them (many moved to Canaan, the informal settlement that emerged in the north of Port-au-Prince—see the introduction). As in the Gujarat reconstruction process that I unpacked in chapter 4, cash subsidies and credits were also provided to affected families. Those who received subsidies were more satisfied than those who received finished homes provided by charities.

Providing resources directly to families not only permits the adaptation of housing solutions to individual needs, opportunities, and expectations, but it distributes risk among stakeholders, including the beneficiaries themselves. As such, there is a transfer of responsibility to those who can make the best choices and who are interested in keeping reconstruction costs low. The Colombian and Gujarat cases show that, when residents manage their own resources, they look for the most affordable materials and components and recycle any structures saved from the disaster. The result is that, with fewer resources, there are fewer housing deficits.

Unfortunately, charities and governments often target the construction of brand new, finished houses, and new neighborhoods in remote locations. Many believe that construction companies can provide better quality and speed up the process through their experienced builders. Officers also worry that users might waste the resources given to them by making unnecessary purchases. In some cases, the actual reason is the desire to facilitate management by transferring efforts to a single contractor. Other times, concerns are closely related to corruption—the possibility of giving handsome contracts to companies that are politically compromised.

While increased freedom to build has worked well for rural residents in Colombia, cities require even more intense interactions between buildings and infrastructure than those in the countryside. The same level of freedom can cause problems when applied to dense areas, so in 2006, then mayor of Facatativá, Colombia, Alvaro Bernal, devised a solution.

GIMME SHELTER, BUT ONLY HALF OF IT

Like many Latin American cities, Facatativá was frequently affected by floods, while hundreds of low-income residents squatted land near rivers. Alvaro Bernal realized that it was necessary to relocate about five hundred of the families living in Factatativá's river shores. He also anticipated that about 1,500 additional homes were needed to reduce housing deficits in the city.

"Building two thousand homes is a major challenge for any city, let alone for a small one of about 140,000 people," Bernal told me in an interview in Facatativá. The mayor didn't have the resources to build complete homes for two thousand families, but he found that he had "enough money to build two thousand half—or unfinished—houses."

Knowing that most low-income residents and slum dwellers in Factatativá were used to building their own houses, Bernal decided that the city could build small core units and give beneficiaries the opportunity to finish their homes later on, when they had the resources required. Bernal understood that, in order to succeed, the new residential development had to be closely connected to the city. He knew that relocating families far away would only cause additional problems and amplify vulnerabilities. Bernal also knew that any initiative would have to include infrastructure, public parks, sport facilities, a school, and a library. But he did not have the funds for all of this. Bernal's administration bought a piece of land that could be integrated with existing residential neighborhoods. He also employed a construction company to deal with infrastructure and the construction of core units. The city started the construction of the school and the library, but there was no money left for other facilities. For this reason, the city left a generous green area intended for sport facilities in the future. "We also prioritized pedestrian paths over automobile roads," affirmed Bernal, "since all families live within a maximum of 100 meters from a vehicle road." This helped minimize investments in infrastructure and provided an open area that could later be appropriated by residents.

Each family eventually received a two-story, 40-square-meter structure, with a backyard where they could add an extension to the house (for a maximum of 80 square meters). When the core units were given to families, both the homes and neighborhood looked like a construction site. The core units were very simple and lacked basic furnishes, such as kitchen cabinets, tiles, paint, bathroom details, and ceilings. I wondered how residents were going to respond to this and feared that they would be offended by these units that didn't even have a bathroom mirror.

But the city helped beneficiaries connect with financial opportunities and subsidies provided by the national government. They also granted direct subsidies for the most vulnerable, notably those members who had to be relocated from the river shores and didn't have the essential credit capacity. Residents also received a construction guide with specifications for building structurally-sound additions to the core units.

The idea paid-off. When I visited the project, six months after the delivery of these core units, residents had already started to work on their expansions. In subsequent visits, I witnessed the enthusiasm with which beneficiaries finally finished and enlarged their core homes. Many residents built stores on the ground level, while others constructed workshops or opened new businesses, such as cafeterias or hair salons. Most home-based businesses in the settlement are, still to this day, run by women, as they can efficiently care for children and the elderly while working from home.

A positive feedback process quickly kicked in: as residents developed economic activities and home-based businesses, their economic conditions improved. This allowed them to invest in enlarging and upgrading the core units, as well as improving their businesses. A few years later, it was clear that residents had not only invested in finishing their homes but also in transforming public spaces. They invested their time and resources in gardens, parks for children, and landscaping.

Although many of the additions failed to adopt the structural standards outlined in the construction guide, the residents altered the original housing façades to personalize their homes. Perhaps most importantly, the core units are safe and the neighborhood keeps a general coherence.

Bernal's leadership was crucial throughout the entire process. And yet, his intervention didn't stop the city from sharing significant decision-making power with the residents. The aim of Bernal's plan was to reduce vulnerabilities and make space for individual choices. This is not the typical delegation of expenses and activities found in the resilience agenda. But Bernal's administration provided a concrete solution to risk reduction that integrated financial institutions, government agencies, private companies, and citizens themselves.

At the end of his mandate, Bernal went on to work with the national government. Other social housing projects were built in Facatativá, but none replicated the idea of half-finished homes. Instead, tons of five-story social housing buildings were rubber stamped in areas close to Bernal's neighborhood.

AN ALTERNATIVE NARRATIVE TO THINK ABOUT RISK AND DISASTERS

The Medellín transformation, the major reconstruction in rural Colombia, and the new neighborhood in Facatativá, did not owe their success to the promotion of innovative solutions in housing developments or to the organization of community meetings and audiences. Even though there were serious concerns about protecting the environment, there was neither a sustainable development narrative nor any resilience rhetoric used to guide the process. Instead of resorting to superficial communications about sustainability, resilience, innovation, and community participation, these initiatives focused on taking care of the most vulnerable citizens and attending to their most pressing needs and expectations.

The process was successful because there were strong institutions behind it—all closely linked to the Colombian state. These institutions were determined in their goals, yet they simultaneously decentralized decision-making power and transferred it to those who could make certain choices: citizens, private companies, and civil society groups.

The transformation in Medellín, the reconstruction of eje cafetero, and the case of Facatativá demonstrate that strong institutions need not be autocratic, centralized, and oppressive—the traits that neoliberal defenders often fear. Nor must they be dangerously driven by ideology. Surely, the Factatativa administration, the EPM, FOREC, and the Federation of Coffee Growers were all motivated by clear objectives but resorted to strategies that respected local traditions and values. Ultimately, the Medellín institutions are built on strong local values and business practices that have not prevented EPM from becoming a powerful international company. This company has activities in many Latin American countries but remains close to the needs and expectations of low-income and vulnerable residents. Bernal had a clear path to follow; he recognized the integral role of decision making among both residents and companies to succeed.

These cases, and the one in Gujarat that I discuss in chapter 4, show that strong institutions can be of a different nature than those that James Scott represents in *Seeing Like a State* (refer to chapter 2) or that economists might otherwise fear. Strong institutions are not necessarily those that design master plans to replace the existing urban fabric and structures with new idealized ones. Unlike the case in the UK (illustrated in chapter 4), strong institutions do not transfer activities and action without transferring decision-making power. Bernal's administration, FOREC, the Federation of Coffee Growers, and the Gujarat Authority had a key objective, but each always showed respect for the neighborhoods and vernacular solutions created by the poor and marginalized. They worked to build upon their foundation and refine the infrastructure, while maintaining social networks, adopting vernacular constructions, and embracing local services and companies that existed.

These institutions are extensions of the power of the state. But they give sufficient space to private companies and economic markets. FOREC and the federation, for instance, gave private companies an opportunity to sell construction components through the housing exhibition in La Tebaida. The Medellín institutions have also partnered with the private sector to conduct innovative projects. But they did not avoid their *own* social responsibility, nor did they try transfer the development of their core mission to free markets. Instead, they acted as leaders, capable of integrating the private sector—*when* and *how* it was necessary.

By guaranteeing that they are properly funded and supported, the state assured that these institutions had specialized, skillful staff. Highly trained professionals in these institutions were able to find appropriate solutions for the most vulnerable. Bernal's administration, FOREC, and the Federation of Coffee Growers acknowledged that the real value lies in helping the most vulnerable. To a certain extent, these have been also the values behind the rapid transformation of Medellín and the reconstruction in Gujarat. They too have been able to understand the role that citizens play—and can further play—to reduce risk and react diligently after disasters.

The Medellín case, the Facatativa example, the rural reconstruction in Colombia, and the Gujarat reconstruction case I examined in chapter 4, are not blueprints for other scenarios. They are the product of contextual characteristics and a certain history, and thus they cannot simply be replicated. But they point us to a narrative that is distinct from sustainable development, green solutions, resilience, and high-tech innovation everywhere.

These cases support a tradition of scholarship that has focused on finding alternative narratives to the improvement of living conditions in the Global South and on decolonizing them from Northern paradigms.[31] As such, they represent an alternative to the other four underlying narratives of progress and risk reduction that I explain in the previous chapters—and that are often associated with the sustainable development ideal.

First, are the common narratives often mobilized by the Left, which focus on the dangers of change caused by notions of progress as seen by the state. Most of these narratives argue that a strong state sooner or later falls in abusive centralization and obsessive ideology, using its power to erase vernacular practices and citizens' differences to control people and resources (refer to the centralized schemes that Scott found in *Seeing Like a State* in chapter 2).

Second, is the frequent narrative often mobilized by the Right that argues that free markets and privatization are the only sustainable options to confront contemporary challenges. This is the neoliberal idea that, in order to avoid the danger of state oppression, more responsibilities and power must be given to free, deregulated markets. In this narrative, profit-seeking companies must be arbitrated by the invisible hand of the capitalist (hopefully green) economy.

Third, is the common narrative in the North that claims that sustainable progress can only be achieved through aid. According to this theory, the poor and marginalized are trapped in vicious cycles of vulnerability-disasters-vulnerability that can only be broken through external intervention. With aid, the poor and marginalized can adopt the best practices and technologies and become greener.

Fourth, is the narrative of both academics and neoliberals that argues that aid is actually a waste of money and extends patterns of domination. Progress is not achieved by planners but by searchers. According to this narrative, progress requires the transfer of responsibilities to the private sector, which is the only sector capable of finding solutions organically and efficiently.

STRONG INSTITUTIONS AND THE REDISTRIBUTION OF PRIVILEGES AND FREEDOMS

In the alternative narrative that I expose here, there are two central ideas to keep in mind.

The first idea is the establishment of proper governance mechanisms, where fair public institutions—an extension of the state—lead and coordinate activities aimed at reducing vulnerabilities within (and not through the replacement of) existing solutions. These institutions set strict regulations, procedures, and collaborative mechanisms to incorporate the private sector. But contrary to neoliberal assumptions, these public institutions do not delegate responsibilities and grandiose ideas to capitalist markets. They assume a long-term commitment to protect the most vulnerable and the environment. They do not search for blank slates; they act as microsurgeons, with a keen eye for detail "on the ground." This allows them to provide support and resources at a microscale, without losing sight of the larger picture.

Strong public institutions frame and constrain the wealthy and powerful, so that they do not abuse their privileges, exploit nature, or pollute it in the process. Here, public institutions discern the contribution of private companies, NGOs, aid, and international agencies, but they assume the leading role in regulating their involvement.

The second idea is the recognition of the agency of citizens and vulnerable people to make decisions. This narrative builds on an expansion of vulnerable citizens' freedoms to act upon their own futures. It implies transferring decision-making power to manage resources, set goals, adopt implementation mechanisms, and certify the continuity of initiatives. This is not a narrative of communities participating in master plans designed by the authorities. It is one in which vulnerable citizens are given space for direct decision making within an overall framework that prioritizes the protection of nature and citizens. It is a narrative of *care, solidarity, and empathy* for the have-nots, the marginalized, and historically excluded; one that recognizes the strengths of all citizens and their capacity to transform their own environment.

Is this wishful thinking? Perhaps. But there is enough evidence in this book that underscores the necessity of challenging certain narratives that have been imposed on us in the name of sustainable development and resilience for responding to risks and disasters.

The problems that we face today require public institutions to stop acting as master plan designers that shape reality as if starting with a blank slate. These institutions will have to work at a microlevel, while maintaining overall coherence and integration. But more crucially, they will have to restrict some of the privileges currently enjoyed (and abused) by the most powerful companies and citizens. Meanwhile, they will have to increase freedoms and spaces for the most vulnerable to act. In other words, a world of unnatural disasters requires a more decisive redistribution of privileges and responsibilities. Additional restrictions are needed to contain and reduce the negative effects often caused by mighty corporations and wealthy populations. Rights and freedoms will have to be awarded to low-income citizens, historically marginalized social groups, and fragile businesses, so that they can make the choices that directly impact their own trajectories.

There are many reasons why it is necessary to restrict the current privileges of wealthier citizens and mighty corporations and companies. Here is a reminder: They disproportionally pollute the atmosphere and produce carbon emissions that put all of us, and the planet, at risk. They consume more than others, generate more waste, and use up more resources. They also tend to get closer to nature in dangerous ways, destroying it in the process. They privatize land, water, and other vital components of the environment. They are often primarily motivated by profits and neglect the traditions, local rituals, place attachment, freedom of action, and other forms of human dignity that are important to vulnerable people. Finally, they have shown an unrestrained propensity to impose their own view of the world on others, displaying a dangerous sense of superiority.

Now, there are reasons why the have-nots, the marginalized citizens, and fragile businesses require additional privileges: they are often less heard, protected, integrated, and supported when it comes to facing risks or reacting after mass destruction. The final chapter reminds us why listening to those who are most affected by risks and disasters matters.

8 Humility

"The Damn Circumstance of Water Everywhere"

This statement quieted the public's concerns. But it also produced other effects perhaps unforeseen by its author. It profoundly altered both the spirit and the operations of the Company. I have but little time remaining; we are told that the ship is about to sail—but I will try to explain.

—Jorge Luis Borges, "The Lottery in Babylon"

THE CITY OF FOREIGN CONSULTANTS

During the dry season, hundreds of women gather once a week to create a temporary market in Pétionville, a neighborhood in the high mountains that surround Port-au-Prince. They meet at dawn and transform a *ravine*, or urban drain, into a fully functioning market. For a few hours, they sell fish, fruit, clothes, rice, second-hand appliances, and anything else you might need. The market extends for almost three hundred meters in a 40 meter wide ravine (see figure 8.1). At noon, the women dismantle their stalls and leave, returning the ravine to its original form.

FIGURE 8.1 A temporary market is set up once a week in a ravine (or urban drain) in Pétion-Ville, Port-au-Prince, Haiti.

Source: Gonzalo Lizarralde, 2014.

This temporary market is perhaps one of the most beautiful spectacles of urban space appropriation in Port-au-Prince. Traders in Haiti convert the ravines and also set up their businesses on sidewalks and streets. Driving through Port-au-Prince, it feels like you're crossing a limitless open market of goods. Yet these improvised markets are, above anything else, a survival strategy for most traders, notably women. Seeing as they do not have the skills required for permanent jobs in companies or the government, many of these women turn to informal economic activities to both increase family income and gain a certain independence from their oppressive husbands.

Most affluent Haitians complain that street vendors invade the sidewalks, roads, and parks. They view the prevalence of informal traders in public space as another manifestation of "anarchy" in the country. Many in the city would like the government to evict traders and reclaim public space. They look back with nostalgia to the days when Champs de Mars and other such public spaces looked like Parisian parks. In a study I conducted with a team from Université de Québec à Montréal (UQAM) in Port-au-Prince from 2015 to 2018, we found that one of the most common narratives regarding informality concerns its chaotic nature. In Haiti, informality is often associated with chaos, and thus, considered a problem to be eradicated.

But street businesses are crucial for low-income families and the economy in general. Plus, selling stuff is just one of the activities that you might encounter on Haitian roads. The streets and sidewalks of the capital city are always busy. People gather and chat right in the roadway, while cars and motorcycles find parking on the sidewalks. Roadways are occupied by animals, and in many cases, children playing soccer, baseball, and basketball. On almost every street of Port-au-Prince, cars and buses compete with animals and pedestrians to use the roads.

When the 2010 Goudougoudou and subsequent measures prompted the emergence of Canaan, charities worried that the settlement would become another slum in Port-au-Prince. Many residents in Canaan rely on informal activities financially, and they depend on underground wells for access to water. Very few locals have cars, and so they require the use public transportation to commute to Port-au-Prince for work and services. Despite the fact that residents have built roads and left spaces for the construction of future parks and infrastructure, authorities and charities remain troubled by the prospect of the settlement's deterioration. They fear that Canaan follows the path of notorious slums such as Cité Soleil or Cité Lajoie.

The American Red Cross has offered health services in Canaan since the early days of its emergence. Years later, the organization decided to expand its role and build public spaces in the area. Between 2015 and 2018, the U.S. organization invested US$21 million to improve Canaan's infrastructure and parks. The institution then partnered with the UCLBP, a newly created Haitian government unit in charge of housing, construction of public buildings, and urban planning. In 2015, the UCLBP revealed its intention to restructure Canaan "in accordance with urban planning standards and principles, and in keeping with the international vision of urban development."[1]

One of the first initiatives put forth by UCLBP and the American Red Cross was to set up a structure of neighborhood committees (*tables de quartier*) with local leaders. This network of community-based organizations served to provide a valuable structure to facilitate the implementation of changes, such as building parks and water wells. In order to produce a plan for the settlement, the Red Cross went on to partner with one of the most respected agencies in urban and housing matters, United Nations Habitat. The UN had already established a series of planning documents and reconstruction endeavors in Port-au-Prince. Most of them were successful in proposing a "microsurgery" approach, as well as building solutions in infrastructure and urban space without demolishing existing constructions in slums. But UN Habitat sought to create a master plan in Canaan that would follow sustainable development principles, including a comprehensive participatory process of urban planning. To do this, UN Habitat set up a series of participatory workshops (called *charrettes*) with local leaders and representatives of several public and private organizations.

In November 2015, I attended one of those participatory workshops. At first, I was pleased to see that several local leaders and professionals were invited to participate. Through them, my coresearcher Anne-Marie Petter and I learned that there are more than 600 kilometers of dirt roads in Canaan, all designed, built, and maintained by residents themselves. If anything, there were too many roads in the settlement, as few Canaanites actually own cars or bikes. During the workshop, we divided into five or six groups to exchange ideas about the construction of future roads and parks. It soon became clear, however, that the most pressing problems in Canaan involve deforestation, erosion, and lack of water and sanitation. Longer droughts and reductions in precipitation, both exacerbated by climate change, are making these issues unbearable for Canaanites. Residents proposed to plant new trees, build more gardens, develop solutions for agriculture, and initiate local industries around harvesting Moringa—a popular plant in the region used for healthcare purposes.

During the workshop, I found myself at a discussion table with Natasha Belleville, a young woman I had met a few days earlier. When I met Natasha, I had the impression that she had more ideas for Canaan than time to explain them. As a resident of Croix-des-Bouquets, a commune that is part of metropolitan Port-au-Prince, Natasha manages a program to keep children busy after school. She also works with victims of domestic violence and regularly participates in activities coordinated by international NGOs in Haiti. Natasha has a profound knowledge of slums, which makes her a crucial asset for NGOs interested in implementing new programs in informal settlements. During the UN Habitat workshop, she wanted everybody to focus on environmental degradation in the region.

As the event wrapped up, the UN Habitat representative summarized the results. We learned a little bit about what the participants had discussed at their respective tables throughout the session. But he also had another message to deliver: urban changes in Canaan had to comply with the UN Habitat standards for city streets and roads. His announcement was backed by a flashy Power Point presentation displaying sections and street plans similar to those you would find in suburban America. While he also made passing references to agriculture, Moringa, reforestation, and gardens, the UN representative dedicated more time to exposing a new vision for city planning, the organization of urban systems, and the importance of developing a "proper" network of roads in Canaan.

I thought that these were only general references meant to spark discussion, and that later, the outline would integrate the proposals that had been conceived by Natasha, other local leaders, and professionals. But weeks later, when UN Habitat revealed its final plan for Canaan, we discovered that the Power Point presented in November was not a *general guideline* for urban design—it was *the plan* for Canaan. The proposal did not reference the integration of economic and agricultural activities, how to contain deforestation, or to combat erosion. Canaan, instead, became the subject of a UN-sanctioned plan for large paved roads, with sidewalks and lighting that would emulate an American suburb. The street widths were determined according to standards of heavy traffic, without considering the presence of street vendors. UN Habitat reserved the streets for cars, as opposed to the ample activities that typically go on in Port-au-Prince.

The plan concluded that Canaan should comply with international urban planning standards. It was, however, idealistic and disconnected from Haitian urban realities. Not only are there are few cars in Canaan, but the streets proposed by UN consultants disregarded the economic activities that traditionally take place in the urban streets of Haiti. If ever built, the plan would significantly increase paved (and thus impermeable) surfaces in

Canaan. This is likely to accelerate water run-off in the case of heavy rains and reduce the amount of water that percolates on the ground and feeds underground wells. It also brings up the question of who will pay for these new roads. My colleague Anne-Marie Petter, who has since studied Canaan extensively, estimated that "expanding the road network alone, from 600 km to 1,500 km as detailed in the UN Habitat plan, would cost an estimated US$1.5 billion, or 81 percent of Haiti's national budget for 2018–2019."[2]

International consultants of UN Habitat, UNDP, UNDRR, the Rockefeller Foundation, ICLEI, and other international agencies, often argue that the plans and initiatives they put forward are not necessarily theirs, as they partner with local institutions and merely act as consultants in the preparation of reports and policy documents. While it is true that UN Habitat partnered with UCLBP and the American Red Cross to produce a plan for Canaan, it is difficult to argue that the international agency didn't have a strong influence on the solution that was eventually presented. As the scenario of the UN Habitat Power Point elucidates, the urban solution had been selected *before* organizing the participatory workshops.

But, why were international consultants responsible for the development of this plan in the first place? Surely, most public institutions in Haiti suffer from corruption, mismanagement, and cronyism. Politicians and technocrats in the country are often seeking to benefit themselves (and their families) rather that the poor and marginalized. But relying on international institutions to design urban plans does not fix these problems. Besides, there are numerous honest professionals and excellent Haitian architects, engineers, and urban planners in Port-au-Prince. I have met many of them, seen their work, and can attest that they are totally capable of devising the type of plans that Haitians and Canaanites need. Some of them were invited to participate in the UN Habitat workshops, though they would have preferred to be awarded a contract for analysis, diagnosis, and planning. Instead, such contracts were awarded to an international agency that had little knowledge of local realities, problems, and expectations on the ground.

The case of urban planning in Canaan highlights the danger of ignoring local competent professionals while producing disaster reduction and reconstruction solutions. As we shall now see, residents often have the required knowledge of local conditions to think up desirable solutions—ideas that frequently differ from what consultants deem appropriate.

THE MAGIC BALL

Few people think about soccer when looking for solutions to floods, draughts, and the effects of climate change. But Julio Melendez is not like most people. He is fighting climate change and disaster risks in the Colombian city of Yambarí[3]—with a soccer ball as his main weapon.

Julio lives in Los Andes, an informal settlement in Yambarí, an industrial city in Colombia. For years, Julio has been trying to convince authorities and local politicians to build a soccer pitch in his neighborhood. He is sure that soccer can bring substantial urban change and social benefits—and he's close to reaching his goal.

Most families in Yambarí and other low-income neighborhoods in Colombia were displaced from rural areas, in part due to rural violence, unbearable poverty, and social exclusion. Luckily, most rural migrants who move to the city manage to escape poverty within one or two generations. Even with a modest income, families are careful to improve and enlarge their makeshift houses built in the slums. Additional space is used for rent or to conduct home-based economic activities. As a consequence, family income increases, allowing for upgrades and possibly the construction of additional space. Through this positive feedback loop, it is possible to send kids to college or university, so that the following generation of skilled workers avoids poverty. Nonetheless, the future is not so promising for many young Colombians. That's where Julio comes in.

Violence and crime are all too common in Los Andes and other informal settlements. Women fall victim to violence and sexual abuse both within their homes and in the streets. Dozens of teenage girls become pregnant in Yambarí's informal settlements each year, oftentimes taking on the responsibility as single parents. With children to take care of and the need to work to earn money, these girls and teenagers lack the opportunities to study or to find permanent life partners. They struggle to land steady jobs, and unsurprisingly, fail to escape poverty.

Young males, meanwhile, are often tempted to join gangs and partake in organized crime. When I spoke with a policeman in Yambarí, he told me that, years before, most assassins were males between ages twenty and forty years old with access to weapons bought from the black market. "Today," he says, "many are under twenty and use makeshift firearms from the slums." Youngsters from one neighborhood can't go into another one because mafias have created "invisible barriers" in the territory, the policeman explains.

Fortunately, Julio has detected a reassuring pattern. On Sundays, young people from different areas of the city gather to play soccer in a nearby neighborhood. They are proud to be represented in the local league's championship and actively support their team by cheering them on at the soccer pitch each Sunday. During these soccer games, the aggression and violence that characterizes gang culture becomes a friendly competition.

While Julio never advanced to a professional league, he has always been a strong soccer player and joined several lower-level teams. Then, for many years, he belonged to a Colombian guerrilla. After witnessing terrible violence firsthand, he decided to alter his path. Today, he trains children and teenagers on a local soccer team in Los Andes, and he also involves them in environmental activities. He organizes outings to plant trees and clean drains, and he teaches them about the importance of knowing the territory. By telling them about his personal experience in the guerrilla, he tries to convince the children to reject violence and prevent them from joining gangs or criminal groups.

"This soccer ball," he tells us, "is a magic weapon." With it, Julio is able to attract and include youths in sport activities and environmental causes. "Once we have the kids here, we can plant trees, clean the water drains, learn how to manage water, talk about safe sex, and many other things," he explains. Julio is committed to the protection of children and teenagers in his neighborhood and has found a way to gain their trust.

Now Julio wants his neighborhood to have its own football field. He believes that if a football pitch is built in Los Andes, he can take his initiative to new heights: "There is so much we can do to fight climate change and improve people's lives when we capture the attention of children and young men and women."

FRAGILE LEADERSHIP

Although Julio is not alone in his cause, my colleagues from Canada, Colombia, Cuba, Haiti, and Chile, have found that, in a way, he is a rarity. This is because most leaders of social and environmental causes in informal settlements are women. They take action; tend to plants, gardens, and food in slums; they collect, filter, and boil water—all while caring for both children and the elderly.

In Yambarí, one of these local leaders is Ana Tavarez. Once a week, she gathers with other women to plant trees and educate children on the importance of protecting the environment. Ana leads this initiative in one of the most polluted, dangerous, and neglected parts of the country. For more than five years, she campaigned for the construction of parks in informal settlements in Yambarí. It was a difficult task, as she had no political influence and little visibility to voice her concerns. But when she partnered with Adriana López and Owaldo López, two professors of Universidad del Valle (an institution based in Cali), she was able to raise more awareness about the importance of public space in the slums. With the help of committed academics, her cause seized the attention of politicians, donors, and private companies. But just as the project was finally taking off, Ana decided to step down. Her initiative was eventually run by local politicians and the Universidad del Valle team and became a great success.

Unfortunately, Ana's case is not rare. Our study on climate change adaptation in Latin America suggests that local leadership in risk reduction and climate change action is remarkably fragile in informal settlements. While women are usually the ones who initiate change, it is difficult for them to maintain leadership until completion. There are several reasons for this: One is that women like Ana are not interested in navigating the dirty paths of local politics. Ana would rather help her people and protect the environment than play political games. Another reason is that local leaders like Ana often lack the financial literacy for planning and budgeting. In addition, most women in informal settlements do not care for personal recognition—they are rarely interested in the type of leadership required once projects gain popularity and prominence. Ironically, there are almost always men willing to assume that leadership, obtain visibility, and play the political games that lead up to project execution.

Keeping this in mind, our local partners in Latin America, including the NGO Corporación Antioquia Presente, are developing training activities for local leaders in the slums—and for women, in particular. We are currently evaluating the effects of this program to see if a different type of leadership in the region makes a difference when it comes to climate change response.

ARTIFACTS OF DISASTER RISK REDUCTION

A woman in Shanghai sets up a food stall on the sidewalk. She has an indoor store, but the stall attracts more passers-by and increases her revenue. At five o'clock, she removes the stall so that the sidewalk recovers its original state.

Entrepreneurs in Hanoi transform their small apartments into businesses. They sell clothes, cut people's hair, or repair cameras and electronics.

In Montreal, residents post a sign in front of their church: "Do not throw garbage here. If we see garbage around the church, you will be fined $100." I suspect that, while the regulation is presented with sufficient authority, most citizens don't know whether the threat is actually enforced.

These are all cities in the making in times of unnatural disasters.[4] Their solutions fall into what Bernard Rudofsky, a Czech American writer, calls "non-pedigreed" design;[5] often described as "informal," "self-help," or "community-based." In rich cities, they are known as "do-it-yourself," "local," "community-driven," or "citizen-led." They all have something in common: they do not follow the plans and standards prescribed by the authorities. Some are individual endeavors (for example, a woman builds a canal to prevent her house from flooding), whereas others are collective (a community group transforms the neighborhood park). My colleagues and I refer to them as "artifacts of disaster risk creation." Generally speaking, these correspond to objects, rituals, practices, events, and spaces that have social value, and which, in the face of risk, help connect people, meanings, and agency in a given space.

Most often, they are *bottom-up* activities—as opposed to top-down ones led by authorities, NGOs, and corporations. To be sure, neither artifacts of risk reduction nor informal bottom-up solutions are illegal or illegitimate. As a matter of fact, it is estimated that about half of urban development enterprises in cities in the Global South occur informally or through bottom-up interventions. About one-quarter of the world's urbanites live in what are known as *comunas*, *favelas*, *tugurios*, slums, townships, shanty towns, and other settlements. This type of urban development is seen as both a cause and consequence of the informal economy of developing countries,[6] largely composed of nonregistered microbusinesses like the ones women manage in the Pétionville market.[7]

In my work, I have often encountered two reactions to these bottom-up solutions: There are those that demonize them, associating them with urban chaos, disorder, and opportunism.[8] Many wealthy Haitians, for instance, refer to bottom-up solutions as "anarchic," while Colombian authorities label informal traders as "illegal occupants" of public space. The authorities in South Africa even reduce backyard dwellings built by low-income residents in the townships to "illegal constructions." On the other hand, there are those who romanticize informality. They view these solutions as heroic and creative achievements by resourceful entrepreneurs struggling to survive in a context of poverty, marginalization, and exclusion; solutions that not only emerge in response to urgent needs but are also celebrated for their innovation.

Of course, informed scholars and professionals adopt a more balanced view. They reject the classification of neighborhoods and economic activities as either formal or informal. Instead, they consider the city and its economy as a wide spectrum that ranges between total control by formal authorities and residents in control. They also reject the idea that professional solutions are inherently superior to those developed by ordinary citizens. They recognize that bottom-up solutions have several limitations for responding to the needs and aspirations of the most vulnerable, while acknowledging the ways that formal and informal practices overlap or complement one another. Vulnerable people rarely engage in this debate. I have never heard a low-income resident describe their neighborhood or business as "informal." Citizens simply make changes to front yards, façades, streets, parks, plazas, and sidewalks without distinguishing between formality and informality. People also seek out business opportunities without bothering to categorize them as formal or informal.

Cities have always been modified without construction permits or approved plans. Various spaces, in rich and poor countries alike, are harnessed in unexpected ways. Some believe that it is precisely the aggregation of individual action over time that constitutes the character, appeal, and beauty of vibrant cities. In the diversity of individual interventions (with or without permits) emerges a sort of organic order. People's interventions in the city contribute to a humanizing form of urban delight that creates the charming effect of historic neighborhoods; a feeling that we almost never find in planned districts or suburbs. But once disaster strikes, we tend to change our tune.

WHEN DISASTER STRIKES, THEY ARE HELD RESPONSIBLE

Authorities and wealthy citizens often blame disaster destruction and risk on bottom-up solutions conducted informally. Many Haitians and foreign observers believe that the destruction triggered by the 2010 earthquake in Port-au-Prince was caused by the poor construction of informal builders. Residents of Pétionville affirm that vendors pollute water, air, and soil by claiming public space for their personal economic activities in makeshift markets.

When a landslide destroyed part of the city of Mocoa, Colombia in 2017, several journalist and "experts" blamed slum dwellers for occupying risk-prone areas close to the rivers. In chapter 2, we saw how *favelados* were accused of deforestation and environmental degradation in the mountains around Rio de Janeiro. Similarly, informal vendors in Latin American and Asian cities are accused of intensifying pollution, traffic, and other urban threats.

When citizens modify human settlements and territories, they partake in a constant battle for space, rights, and opportunities in the city. In this war for space and services, there is a sense that only the toughest, through their individual might or collective effort, will win.[9] But studies show otherwise.

Although bottom-up interventions in the city indeed serve individual goals, some academics assert that these initiatives evolve into collective action. Contrary to prevailing beliefs about individual competition, there is always some form of cooperation and coordination within social groups.

A study conducted by Georgia Cardosi, an Italian researcher based in Canada, demonstrates that in Kibera (Nairobi), one of the largest informal settlements in Africa, residents collaborate on the design of public spaces and amenities.[10] Anne-Marie Petter has found a similar pattern in Canaan, Haiti. While this is not meant to imply that the struggle for space in the city does not exist or that it is not important, it proves that these struggles are not necessarily motivated by greed or selfish gains. In reality, competition coexists with cooperation in the face of risk.

Community-driven design is a common strategy for individuals and social groups to respond to risk (hence our term "artifacts of disaster risk reduction"). Citizens in Brazilian favelas build fences around their houses in an attempt to prevent crime and vandalism. Some residents in Canaan install barriers on the roads to restrict access to their neighborhoods, while people in Nueva Choluteca, Honduras, plant trees in their yards to shield their dwellings from wind or direct sunlight. Most of them also fortify their windows and doors with metal grids to prevent break-ins. In all these cases, bottom-up design is a form of protection fuelled by local knowledge of dangers and individual perceptions of risk.

These bottom-up solutions are comprised of local expertise and insight, often deriving from first-hand accounts that evade authorities. Unlike professional design, citizen- or community-led design responds to the real needs and expectations of residents. For example, the path used by pedestrians who cut across the park (instead of using the square-angled sidewalks) captures a form of internal wisdom, unknown to (or ignored by) landscape designers. By redefining structures and systems, users act upon their knowledge of local situations. This has led some authors to theorize that buildings "learn" over time, as they continue to evolve through modifications made by their occupants.[11] With age, these spaces become recipients of bottom-up knowledge.

Yet, the people of Canaan, along with Medellín's low-income comunas, or Rio de Janeiro's favelas, transform infrastructure for reasons beyond risk reduction and functionality. They do so to express meanings, representations, ambitions, and traditions.[12] They create artifacts of risk reduction (often mixing spaces and events) to conduct local rituals and culturally-meaningful actions. Slum dwellers in South Africa, for instance, transform spaces to convey community values and reflect aspects of daily life, citizenship, and collective action.[13] These designed artifacts embody cultural identity and strengthen the construction of that identity. Changes to houses or public space in low-income neighborhoods signify underlying aspirations—religious, cultural, and historical. Sometimes, low-income residents adopt vernacular styles, architectural forms, and typologies, while also including aesthetic and functional solutions implemented by wealthier social groups.[14] These bottom-up interventions create a sense of belonging as well as geographical and cultural references. From pastiche houses in Lac-Mégantic and roads across Canaan, to home-based stores in Bonteheuwel or community-built schools in Choluteca, there is a consistent aim to fulfill immediate needs while creating attachments in space and time.

The informal sector is the largest industry in the world. It is responsible for about 40 percent of all construction in developing countries and remains the only industry capable of providing affordable solutions for the poorest sectors of society.[15] But it is often neglected and ignored in risk reduction plans and reconstruction efforts. The informal sector is also ignored in sustainable development initiatives. As we saw in previous chapters, sustainable development is based on a powerful alliance of economic and political powers and relies on high-tech solutions, or the transfer of technology from the Global North to the Global South. There is no designated role for people in informal settlements to participate in the sustainability game.

In chapter 2, we saw how disasters are ideal opportunities for authorities to deploy modernization schemes. Given the right opportunity, they enforce their own vision of sustainable development, predicated on factors of legibility, standardization, and simplification. This not only applies to urban planning but to economic policy, social measures, and changes in the occupation of territory. Decision makers deploy paternalistic approaches grounded in their own assessments of risk and on the assumption that they know what citizens need. In the pursuit of sustainability, they apply blanket solutions to heterogeneous individuals and social groups.

Nonetheless, with or without planning and regulations, citizens engage in significant strategies to minimize risk. Individuals and social groups combine competition with collaboration to claim space and appropriate its use. They respond to climate change and disasters by modifying their behaviors and environments. But people who modify their living environments are not always in survival mode, nor do they exclusively seek to achieve functional objectives. Instead, they are in search of meaning, identity, and a sense of belonging. And as much as change is needed, traditions comfort them when confronted by loss.

Understanding how people modify their cities in response to hardships and risk provides an excellent background to an underlying message I have tried to convey in this book: If we want to fix a planet at risk, we need more humility. We must accept that we do not always have the right knowledge, skills, tools, resources, and ideas to help others and prevent damage among people and ecosystems. Overconfidence in our capacity to produce positive change often blinds us to better alternatives. We can avoid some of these mistakes if we recognize that those perceived as vulnerable sometimes have the knowledge and skills to do the right thing. Humility is also crucial to develop empathy and recognize bottom-up solutions that can be supported and scaled up. In my previous book, I called this approach "learning from the poor." That should be amended to "learning from those at risk."

In the final section of this book, I will address final remarks targeted to citizens in general, academics, and practitioners.

FINAL REMARKS AIMED AT CITIZENS IN GENERAL

We have seen in this book that as environmental problems and social challenges become more complex, our actions as consumers and voters do not always match our intentions. It is increasingly necessary to critically reflect on the types of change we want to produce as citizens. As consumers, voters, donors, and activists, we have the power to change the current state of affairs. We have the power to reduce risk and modify disaster responses. But finding ways to produce positive change is not always easy.

As citizens, we are unsure about how much we are actually willing to pay to protect nature. We might be ready to make small efforts, such as banning the use of disposable straws in fast-food restaurants, but as a society, we do not easily agree on legislative matters, such as imposing extra taxes on gas or carbon to reduce emissions. These taxes would arguably have a greater impact on the environment than the plastic straw ban. But these taxes require a greater sacrifice through the direct impact on our wallets.[16] We all want to protect nature, but as we become wealthier, we also want to enjoy it, ideally with no one else around. So we move to the outskirts of cities where we can enjoy natural features and inspiring landscapes, which in turn, contributes to urban sprawl. We want to reduce consumerism too, yet we direct this conviction toward the unnecessary purchases that others make—not our own "important" purchases.[17] We are quicker to condemn their consumption than we are our own.

We are also unclear about how much power we want professionals and authorities to have over our own security. We tend to believe that it is fine to ban new houses from being built in flood-prone areas, but we are less sure about forcing people who *already* live in flood-prone areas to abandon their homes. Similarly, most citizens approve increasing police presence in the city to reduce the threat of terror attacks, while treating the use of security cameras in residential areas as invasive of personal privacy. We are uncertain about what warrants protection in particular social groups. We might agree that regulation is critical for protecting architectural heritage from developers' bulldozers in historic districts—but what about saving mom-and-pop shops from ambitious urban renewal projects, especially if they would benefit the majority of local residents?

The role of change through planning and regulation also puzzles people—but for different reasons. Some think there is too much of a protective approach, and others contend that there is not enough.

We have never been wealthier in the history of humankind. And yet, we have put our faith in sustainability, resilience, and other solutions that leave the poor and informal sectors behind. Given the current state of environmental degradation, should we enter a phase of economic degrowth? Perhaps. But again, it is not degrowth that will reduce risks and disasters. In order to have an impact on disaster-risk reduction, a degrowth economy needs to alleviate the vulnerabilities of the poor and marginalized. As Isabelle Anguelovski says, "Consuming and producing less is not enough per se. The 'less' needs to be distributed more equally, with people controlling production processes so that cities and rural spaces become more equal."[18]

Instead of seeing disasters as something for someone else to fix, we have to start viewing them as the result of our collective decisions. We need to challenge some common premises.

Most of us, for example, are infuriated when we read that disaster victims have not received aid in the months or years following a disaster. This frustration is understandable. The images of homeless children make us angry. We feel impotent when we see people living in extreme poverty. But this anger and frustration sometimes lead us to make poor choices, such as voting for populist politicians or giving money to charities that promise rapid action but rarely produce long-term value. Another example is our approach to urban density. As our societies grow wealthier, we believe we can live closer to nature and retreat from polluted and congested urban districts. We move to suburbs and low-density areas close to lakes, rivers, woods, and beaches. However, in doing so, we are not only increasing pollution and endangering the environment—we are multiplying our chances of experiencing floods, fires, and other hazards.

This recommendation comes with a caveat, though. We must realize that there is no one-size-fits-all solution. Increasing urban density is generally vital for minimizing human impacts on the environment while providing adequate services and infrastructure to the majority of citizens. This is true for a large majority of cities in the United States, Canada, and many other places where suburbia has taken hold. But density also brings about its own set of problems. Cities near the ocean are expected to suffer from sea-level rise in the very near future. Without significant transformation, their compact neighborhoods are likely to be the epicenters of mass destruction.

This latest point leads me to another problem. In the face of climate change and globalization, some of us increasingly believe "we're all in the same boat." Self-identification as global citizens is a noble thing, but assuming that we all have the same problems often leads us to make wrong choices. We adopt policies and ideas without properly contextualizing them.

Then, there is fear. Those on the losing end of today's unequal distribution of wealth and opportunities are increasingly angry—and that anger helps elect politicians who resort to the politics of fear to advance their own agendas. Surely, risk reduction and response is about politics. We can't help the poor, the marginalized, and others living in danger without embracing politics. But we must be more selective of the type of leaders we chose to guide us towards risk reduction. Is this a matter of choosing between liberal and conservative leaders? Left or Right? Not exclusively. We should not forget that policies that have resulted in disaster risk creation can be found in all places of the political spectrum. Besides, manipulation of the sustainability, resilience, participation, and innovation rhetoric is not exclusive to one particular political party.

This book is an invitation to challenge our own decisions, from where we build our homes to where we donate our money. It is an invitation to distinguish superficial solutions from real ones. We must distinguish the politicians and consultants who promise short-term solutions and resort to greenwashing from those who are committed to protecting nature and vulnerable people across borders. We must distinguish the dangers that some politicians convince us are looming from the actual dangers we face. We must distinguish the charities that merely produce band-aid solutions from those that empower structural change. We must distinguish the products and solutions that simply look green from the ones that actually reduce impacts on ecosystems.

Our way of living affects the environment. We need to evaluate how much risk our lifestyles produce. In other words, we need to be more critical about where and how we live and vote better, donate more wisely, and consume less. If we want real change, we need more empathy, care, compassion, and commitment.

FINAL REMARKS AIMED AT STUDENTS AND RESEARCHERS

For the past twenty years, I have studied disasters, their causes, and responses to them. During that time, I have witnessed three problems that I tried to illustrate in this book.

The First Problem: The Disconnect Between Ideas and Reality

There is no doubt that since the 1980s, social sciences have made significant contributions to our understanding of risks, disasters, and reconstruction. Human and political ecology, geography, urban studies, architecture, development, and political science have all provided pertinent ideas to disaster studies. Project management, international relations, and other subjects of investigation have also explored specific areas that are crucial to this field. But over time, their contributions have aged and concepts have become abstracted. With abstraction arises blunt generalizations, higher levels of intellection, and a subsequent disconnect from the realities that concepts try to explain.

The ideas of vulnerability, resilience, informality, adaptation, and capacities, originated as ideas to unpack phenomena found "on the ground." They were conceived as practical terms to describe processes, outputs, and mechanisms of response that have been embedded in specific socioeconomic contexts. But once scholars found them to be thought-provoking, they started to study these concepts as independent subjects of inquiry—eventually detaching them from the stories, societies, situations, and individuals that they were trying to explain. Once they grew in popularity, a wide range of decision makers, consultants, and politicians began to apply these terms in different contexts until they lost their purpose and meaning.

Students often ask me whether I am against sustainable development, innovation, participation, the theory of vulnerability, the PAR model, or the resilience approach. I tell them that I am not. These ideas were generated in specific contexts where they proved to be useful. But as they gained momentum, they were investigated in the way that isolated viruses are studied in an aseptic lab. Scholars should be concerned by the rapid and indiscriminate adoption of concepts that become abstracted, disconnected from reality, and at times, meaningless.

The Second Problem: From Explanations to Normative Approaches

Abstract ideas such as sustainability, vulnerability, adaptive capacities, resilience, and participation have become the basis of international guidelines and frameworks promoted by UN agencies, international urban consultants, transnational charities, multinational consultants, and other mammoths of the disaster industry. But as these concepts are increasingly employed, they are also depoliticized. They have been promoted as virtuous, while failing to address their limits, cascading effects, blind spots, and other unintended consequences. Most worryingly is that the concepts linked to sustainability and resilience, which emerged as ideas to explain reality, have become normative frameworks. It is no longer about explaining phenomena found on the ground; it is about making sure that people, communities, cities, and enterprises follow the direction envisioned by global institutions. As such, sustainability, resilience, and other similar ideas have become globalized, prefabricated frameworks of ethical reasoning. As increasing numbers of practitioners rely on them for decision making, they fail to produce contextualized solutions and responses to the specific needs and expectations of individuals and social groups. As academics, we cannot condone these superficial approaches. We must be more critical of manipulation, indiscriminate generalizations, and blunt oversights.

The Third Problem: The Disconnect Between Research and Practice

The fact that several scholars have adopted narrow, simplistic approaches to risk and disasters does not mean that we all have. There are numerous academics promoting decolonization, feminist perspectives, politically-relevant observations of economic systems, ethical approaches to risk management, and radical criticism of superficial renderings of risk, disasters, and climate change. But the issue is that these ideas do not have the same impact among decision- and policymakers. They receive less attention from economists, urban consultants, and professionals in architecture, urban planning, design, and engineering. Many of the pertinent discoveries and perspectives encircling disaster risk reduction can't be easily generalized, and so, require complex analysis. They rely on a thorough understanding of historic, cultural, and socioeconomic contexts, which take effort to translate into the types of guidelines and checklists that practitioners can easily sell in corporate PowerPoint presentations and TED Talks.

We must also embrace multidisciplinary studies. For decades, disaster studies from the perspective of social sciences have focused on impacts on people, whereas ecologists, biologists and other scientists have studied how humans impact flora, fauna, and ecosystems. We must now realize that protecting the environment and ensuring the safety of human beings are no longer two different things. Disaster studies must embrace the problem of loss of biodiversity and the analysis of how human beings are destroying nature and causing disasters not only among people but also among plants and animals. We must move away from an anthropocentric view of disasters and start seeing how we (humans) have become an unprecedented disaster that destroys ecosystems and alters landscapes in unprecedented ways.

We must recognize that we can no longer depend on sustainability, resilience, and other similar frameworks to guide our actions. What nature and humans need is adapted judgement and decisions made on empirical evidence on the ground—not rooted in hollow slogans or buzzwords that help clear our consciences but carry little value.

Today, we need better explanations of risk, climate change effects, and disasters. As academics, we must not blindly reproduce popular narratives. We need to challenge them, produce knowledge that helps find new ones, and pay attention to ideas and concepts that can be found in the margins of disaster scholarship. These explanations should not be generalized or globalized. They must stem from local conditions and narratives to reflect local values and reinforce an undeniable commitment to redressing social injustices and fixing a planet at risk. This implies a "human approach" to disaster research and action, based on principles of social and environmental justice.

If we follow this path, there are promising research agendas to explore ahead of us: the connections between people and their territories, risk and disaster attitudes (rather than capacities), the role of people's emotions in disaster risk response, local meanings in disaster perception and reconstruction, understanding of local values in disaster reduction and mitigation, artifacts of disaster risk reduction, and feminist points of view and framings. These agendas do not require embracing the empty discourse of sustainability, resilience, innovation, and participation. What they require is new ideas and attention to marginalized voices. This is the main challenge of disaster studies today.

FINAL REMARKS AIMED AT PROFESSIONALS AND DECISION MAKERS: THE ETHICS OF DISASTER RISK REDUCTION

There is an urgent need for an ethics approach to risk and disaster response. In this book, I have called for thorough, constant, and transparent evaluation of our own actions. I am not calling for a deontological code of conduct in humanitarian action, nor is my objective here to draw on the principles of any deontological approach to disaster studies and practice. But we need to reevaluate our actions as professionals and decision makers. We must recognize the relevance and value of bottom-up decision making and local knowledge and narratives of risk, climate change, and disasters. It is increasingly necessary that we decolonize our language about development, risk, vulnerabilities, innovation, participation, progress, and social change. Policy and programs must be built on a better understanding of the relationships between people and their territories, bottom-up solutions, informal responses to risk, common artifacts of disaster risk reduction, and people's aspirations, emotions, and values in relation to risk, destruction, and development.

As professionals or highly educated individuals, we must recognize that we rarely have the knowledge and skills to help those who need help. We need to pay attention to what low-income and marginalized people do and have to say. We also need to study more, conduct research, and constantly question our previous knowledge and premises. Finally, we must recognize the importance of listening more attentively to those on the losing end of development.

Recognizing that we don't know the answers to reduce risk, solve poverty, deal with racism, or face marginalization is a crucial step to produce positive change. We are rarely equipped to work in contexts of vulnerability, poverty, and exclusion. We must develop new skills, challenge the pertinence of the ones we have, and cultivate new competencies that can help us help others. There are plenty of useful things we can do to reduce risk and help rebuild after destruction, but they are rarely the ones that we can do with our regular professional skill set. Thus, the need to enlarge our toolkit and constantly question the value and pertinence of our own knowledge and actions.

We must celebrate progress in disaster risk reduction worldwide. We must recognize that, in general, humans suffer less today than previous generations from social injustices, natural hazards, and other threats. But we must also be wary of narratives of social progress that dismiss, hide, or underestimate the suffering of minorities, certain social groups, and those who are often less heard. We can't understand progress in disaster risk reduction without understanding the struggles of those who have not benefitted from it. The value of our society should not be measured by the progress of the privileged but by the way we treat those who have not enjoyed those privileges.

When it comes to helping the poor, we all make mistakes, even when we act in good faith. So, as professionals, we must be wary of uncontested ideas of development that gain popularity and become normative approaches to apply at large. We must constantly question the value of those ideas, challenge their premises, and investigate their real impacts and secondary effects on people and territories.

In this book, I showed several examples of mistakes that involved decision makers' corruption, neglect, racism, and sense of superiority. Some decisions could be traced to greed and the power of international businesses, geopolitics, colonialism, and imperialism. But in many cases, there were other causes at play, ranging from decision makers' inexperience, to inattention, to practitioners' intent to oversimplify complex problems. To be fair, working in the disaster field is very difficult. Most decision makers in this sector work in good faith within complex institutions whose trajectories are difficult to change. Companies, charities, municipalities, aid agencies, international banks, government bodies, consulting firms, and NGOs are complex systems that respond to market logic, industry practices, funding mechanisms, and other factors that determine how things are done. It is sometimes impossible for well-intentioned individuals to change the course of institutions' trajectories, even when pedaling very hard in a different (better) direction. The market, legal, administrative, and financial forces that determine decisions are at times simply too strong and entrenched. They often lead to actions that betray the values of those pedaling from within the mighty machine. It is, of course, difficult to condemn wrong actions and decisions when one's job is at stake. But condoning mistakes and dubious decisions also perpetuate social and environmental injustices.

I hope that practitioners who find themselves in the difficult situation of denouncing their employers or other mighty organizations will be able to rely on informed judgment based on empathy, scientific knowledge, and the understanding of human stories like the ones I have exposed here. Finally, I hope that this book will provide them with arguments to make the right decisions for the sake of love of their people.

In 1943, Cuban writer Virgilio Piñera wrote about what it means to live in a tropical, poor island in the Caribbean. The poem was titled "*La isla en peso*" ("The Weight of the Island"). But it became known for its first line: "The damn circumstance of water everywhere." The poem encapsulates an islander's thoughts on water, isolation, poverty, sex, national pride, and love. The character reflects on his own condition: "If I didn't think the water surrounded me like a cancer," he writes, "I could have slept easy." But he also reflects on the vulnerable people around him and their daily struggles. "As they go down, people feel how water surrounds them," and the mighty water (the beast) "hits them in the back," until they realize the weight of their space—"the weight of an island in the love of its people."

New narratives about risk and Goudougoudous must perhaps carry the weight of their territories in the love of their people.

Notes

Introduction

1. A subheading inspired by Juan Gabriel Vásquez's book *El ruido de las cosas al caer.*
2. The Canadian Disaster Resilience and Sustainable Reconstruction Research Alliance (Oeuvre Durable for its acronym in French).
3. Some local analysts criticized the short-term view espoused in many of these planning documents. See, for instance, "Reconstruction of the City Center? Or Public Buildings," Ayiti Kale Je/Haiti Grassroots Watch, September 24, 2012, http://haiti-grassrootswatch.squarespace.com/20fr.
4. Malcomn Reading Consultants, *Building Back Better Communities Port-Au-Prince, Haiti: Request for Proposal on Behalf of the Government of the Republic of Haiti* (London: Malcomn Reading Consultants, 2010), 1.
5. Bill Clinton, "Our Commitment to Haiti," *Innovations: Technology, Governance, Globalization*, special edition, *Annual Meeting of the Clinton Global Initiative*, no. 2010 (2010): 3–5, at 4.
6. Bracken Hendricks et al., "Green Reconstruction: Laying a Firm Foundation for Haiti's Recovery," *Innovations: Technology, Governance, Globalization*, special edition, *Annual Meeting of the Clinton Global Initiative*, no. 2010 (2010): 92–105, at 94.
7. As stated on the Foundation's website: The Prince's Foundation, November 18, 2019, https://princes-foundation.org/.
8. The full report can be found here: UN Economic and Social Council, "Less Than 2 Per Cent of Promised Reconstruction Aid for Quake-Devastated Haiti Delivered, Haitian Government Envoy Tells Economic and Social Council," July 13, 2010, https://www.un.org/press/en/2010/ecosoc6441.doc.htm.

9. Find the whole Haiti report here: *Haiti Relief and Reconstruction Watch* (blog), Center for Economic and Policy Research, http://cepr.net/blogs/haiti-relief-and-reconstruction-watch/.
10. See Mayor Jason's letter in the Haitian journal *Le Nouvelliste*: Muscadin Jean-Yves Jason, "Poursuivi, Jason sort de son silence," *Le Nouvelliste* (Port-au-Prince), June 27, 2012, https://lenouvelliste.com/public/article/106557/poursuivi-jason-sort-de-son-silence.
11. Ricardo Lambert, "Peyi lòk, 3,7 millions de personnes sont en situation d'insécurité alimentaire aiguë," *Le Nouvelliste* (Port-au-Prince), November 6, 2019.
12. Jean Daniel Sénat, "Quand Jovenel Moïse compare instabilité politique, 'peyi lòk' et le séisme de 2010," *Le Nouvelliste* (Port-au-Prince), January 13, 2020.
13. Robenson Geffard, "Rasire Aity, pour la refondation d'Haïti," *Le Nouvelliste* (Port-au-Prince), February 4, 2019, https://lenouvelliste.com/m/public/index.php/article/197829/rasire-ayiti-pour-la-refondation-dhaiti, 2019.
14. Kasia Mika, *Disasters, Vulnerability, and Narratives: Writing Haiti's Futures* (New York: Routledge, 2018).
15. Danio Darius, "L'opposition s'unit et se donne une alternative pour la refondation de l'État à Mirebalais," *Le Nouvelliste* (Port-au-Prince), September 9, 2019.
16. Organisation des femmes haïtiennes, "Lettre ouverte d'organisations de femmes aux dirigeants des trois pouvoirs de l'État haïtien et aux leaders de l'opposition," *Le Nouvelliste* (Port-au-Prince), February 22, 2019.
17. Jean-Louis Le Touzet, "Haïti: 'L'habitat, c'est le chaos et l'anarchie,' " *Le Nouvelliste* (Port-au-Prince), January 21, 2015, https://lenouvelliste.com/article/140559/haiti-lhabitat-cest-le-chaos-et-lanarchie.
18. See a journalist's report at Joe Mozingo, "Sean Penn's Hands-On Aid for Haiti Quake Victims an Earlier Sign of His Risk-Taking," *Los Angeles Times*, January 10, 2016, https://www.latimes.com/local/lanow/la-fg-sean-penn-haiti-20160110-story.html.
19. See, for instance, a report on displacement here: "Haiti," Internal Displacement Monitoring Centre (IDMC), http://www.internal-displacement.org/countries/haiti.
20. Munich Re, *Natural Catastrophes 2017, Topics Geo* (Munich: Munich Re, 2017).
21. Find a pertinent article here: "Weather-Related Disasters Are Increasing," *Economist*, August 29, 2017, https://www.economist.com/graphic-detail/2017/08/29/weather-related-disasters-are-increasing.
22. Internal Displacement Monitoring Center, *2019 Global Report on Internal Displacement* (Geneva, Switzerland: Internal Displacement Monitoring Centre, 2019).
23. Find the table in the database: Hannah Ritchie and Max Roser, "Natural Disasters," Our World in Data, revised 2019, https://ourworldindata.org/natural-disasters.
24. Stephen Leahy, "Hidden Costs of Climate Change Running Hundreds of Billions a Year," *National Geographic*, September 27, 2017, https://news.nationalgeographic.com/2017/09/climate-change-costs-us-economy-billions-report/.
25. "California's Wildfires and the New Abnormal," *Economist*, November 17, 2018, https://www.economist.com/united-states/2018/11/17/californias-wildfires-and-the-new-abnormal.
26. World Health Organization, *Quantitative Risk Assessment of the Effects of Climate Change in Selected Causes of Death 2030s and 2050s* (Geneva, Switzerland: World Health Organization, 2014).

27. Gro Harlem Brundtland, *Report of the World Commission on Environment and Development: "Our Common Future"* (New York: United Nations, 1987).
28. United Nations Office for Disaster Risk Reduction, *Global Assessment Report on Disaster Risk Reduction* (Geneva, Switzerland: United Nations Office for Disaster Risk Reduction, 2019).
29. Ritchie Hannah and Max Roser, "CO_2 and Greenhouse Gas Emissions," Our World in Data, May 2017, revised August 2020, https://ourworldindata.org/co2-and-other-greenhouse-gas-emissions.
30. Find the report here: Rebecca Lindsey and LuAnn Dahlman, "Climate Change: Global Temperature," Climate.gov, August 14, 2020, https://www.climate.gov/news-features/understanding-climate/climate-change-global-temperature.
31. World Wide Fund, *Living Planet Report—2018: Aiming Higher*, ed. M. Grooten and R. E. Almond (Gland, Switzerland: WWF, 2018).
32. "Ocean Deoxygenation," International Union for Conservation of Nature, https://www.iucn.org/theme/marine-and-polar/our-work/climate-change-and-oceans/ocean-deoxygenation.
33. David Wallace-Wells, *The Uninhabitable Earth: Life After Warming* (New York: Tim Duggan Books, 2019).
34. Information obtained from: "Facts and Statistics: Hurricanes," Insurance Information Institute, https://www.iii.org/fact-statistic/facts-statistics-hurricanes.
35. Some authors use the term "heat injustice" to explain the relationship between heat-related deaths and injuries and socioeconomic differences between people.
36. Naomi Klein, *The Shock Doctrine: The Rise of Disaster Capitalism* (New York: Macmillan, 2007); Naomi Klein, *This Changes Everything: Capitalism vs. the Climate* (Toronto: Simon and Schuster, 2015); Naomi Klein, *On Fire: The (Burning) Case for a Green New Deal* (Toronto: Simon & Schuster, 2019);
37. David Harvey, *A Brief History of Neoliberalism* (New York: Oxford University Press, 2007). Quotes are from pages 2 and 3.
38. Valeria Guarneros-Meza and Mike Geddes, "Local Governance and Participation Under Neoliberalism: Comparative Perspectives," *International Journal of Urban and Regional Research* 34, no. 1 (2010): 115–129.
39. Thomas Perreault and Patricia Martin, "Geographies of Neoliberalism in Latin America," *Environment and Planning A: Economy and Space* 37, no. 2 (2005); and John Gledhill, "Neoliberalism," in *A Companion to the Anthropology of Politics*, ed. David Nugent and Joan Vincent (Victoria, Australia: Blackwell, 2004), 332–348.
40. Ilan Kelman et al., eds., *The Routledge Handbook of Disaster Risk Reduction Including Climate Change Adaptation* (London: Routledge, 2017).
41. For other sources dealing with depoliticising factors in disaster-risk reduction see: Sijla Klepp and Libertad Chavez-Rodriguez, eds., *A Critical Approach to Climate Change Adaptation: Routledge Advances in Climate Change Research.* (New York: Routledge, 2018).
42. Portraying women as the natural guardians of ecosystems and nature can be both paternalistic and misleading. But here I refer to research results, not opinions about the role of women in disaster risk reduction. In fact, I refer to results from our

ADAPTO project (2017–2021) that found that most local leaders conducting disaster risk reduction initiatives in informal settlements in Latin America and the Caribbean are women.

1. Causes: "Disasters Happen for a Reason"

1. See, for instance, a *New York Times* report here: Miriam Jordan, "This Isn't the First Migrant Caravan to Approach the U.S. What Happened to the Last One?," *New York Times*, October 23, 2018, https://www.nytimes.com/2018/10/23/us/migrant-caravan-border.html.
2. See, for instance, this press release: "Over 7,000-Strong, the Migrant Caravan Headed for the US Pushes On," CNBC, October 23, 2018, https://www.cnbc.com/2018/10/23/over-7000-strong-the-migrant-caravan-headed-for-the-us-pushes-on.html.
3. Michelle Ye Hee Lee, "Donald Trump's False Comments Connecting Mexican Immigrants and Crime," *Washington Post*, July 8, 2015, https://www.washingtonpost.com/news/fact-checker/wp/2015/07/08/donald-trumps-false-comments-connecting-mexican-immigrants-and-crime/.
4. See details of the caravan at Nina Strochlic, "This Is What Happens When the Migrant Caravan Comes to Town," *National Geographic*, November 20, 2018, https://www.nationalgeographic.com/culture/2018/11/migrants-continued-chapter-3/.
5. Miriam Jordan and Caitlin Dickerson, "U.S. Continues to Separate Migrant Families Despite Rollback of Policy," *New York Times*, March 9, 2019, https://www.nytimes.com/2019/03/09/us/migrant-family-separations-border.html.
6. Kirk Semple, "Migration Surge from Central America Was Spurred, in Part, by Mexican Policies," *New York Times*, April 1, 2019, https://www.nytimes.com/2019/04/01/world/americas/mexico-migration-border.html.
7. Pan American Sanitary Bureau—Regional Office of the World Health Organization, *A World Safe from Natural Disasters* (Washington, D.C.: World Health Organization, 1994).
8. Céline Charvériat and Inter-American Development Bank, *Natural Disasters in Latin America and the Caribbean: An Overview of Risk* (Washington, D.C.: Inter-American Development Bank, 2000).
9. Timothy S. Thomas et al., *Climate Change and Agriculture in Central America and the Andean Region. Project Note* (Washington, D.C..: International Food Policy Research Institute—IFPRI, 2018).
10. Mapplecroft and CAF Development Bank, *Vulnerability Index to Climate Change in the Latin American and Caribbean Region* (International CAF Development Bank, 2014).
11. United States Bureau of Citizenship and Immigration Services, "El Salvador: Information on Recovery and Reconstruction since the Earthquakes of 2001," Refworld, May 21, 2002, https://www.refworld.org/docid/3dec96354.html.
12. Gonzalo Lizarralde, "Organisational System and Performance of Post-Disaster Reconstruction Projects" (PhD thesis, Université de Montréal, 2004), 460.

13. Michael Shifter, *Countering Criminal Violence in Central America* (New York: Council on Foreign Relations—CFR, 2012).
14. Adán Quan, "Through the Looking Glass: U.S. Aid to El Salvador and the Politics of National Identity," *American Ethnologist* 32, no. 2 (2005): 276–293.
15. Tim Golden, "Salvadorans Sign Treaty to End the War," *New York Times*, January 17, 1992, https://www.nytimes.com/1992/01/17/world/salvadorans-sign-treaty-to-end-the-war.html.
16. Quan, "Through the Looking Glass."
17. Thomas Perreault and Patricia Martin, *Geographies of Neoliberalism in Latin America* (London: Sage, 2005).
18. Some of the best contributions in this subject include articles written by Ben Wisner, Mario Lungo, Ricardo Castellanos, and Lidia Salamanca.
19. Perreault and Martin, *Geographies of Neoliberalism.*
20. Carlos Arze and Tom Kruse, "The Consequences of Neoliberal Reform," *NACLA Report on the Americas* 38, no. 3 (2004): 23–28.
21. Ben Wisner, "Risk and the Neoliberal State: Why Post-Mitch Lessons Didn't Reduce El Salvador's Earthquake Losses," *Disasters* 25, no. 3 (2001): 251–268.
22. Gonzalo Lizarralde, "The Challenge of Low-Cost Housing for Disaster Prevention in Small Municipalities," in *4th International i-Rec Conference 2008. Building Resilience: Achieving Effective Post-Disaster Reconstruction* (Christchurch, New Zealand: i-Rec, 2008), electronic publication.
23. Gonzalo Lizarralde, *The Invisible Houses: Rethinking and Designing Low-Cost Housing in Developing Countries* (London: Routledge, 2014).
24. Lizarralde, "The Challenge of Low-Cost Housing."
25. Kelly Hallman et al., "Childcare, Mothers' Work, and Earnings: Findings from the Urban Slums of Guatemala City" (working paper no. 165: Population Council, 2002).
26. Dennis Rodgers, "Slum Wars of the 21st Century: Gangs, Mano Dura and the New Urban Geography of Conflict in Central America," *Development and Change* 40, no. 5 (2009): 949–976.
27. See the graph here: "The World's Most Dangerous Cities," *Economist*, March 31, 2017, https://www.economist.com/graphic-detail/2017/03/31/the-worlds-most-dangerous-cities.
28. Virginia Garrard-Burnett and Virginia Garrard, *Terror in the Land of the Holy Spirit: Guatemala Under General Efrain Rios 1982–1983* (Oxford: Oxford University Press, 2010).
29. Rubiana Chamarbagwala and Hilcías E Morán, "The Human Capital Consequences of Civil War: Evidence from Guatemala," *Journal of Development Economics* 94, no. 1 (2011): 41–61.
30. Peter J Meyer, *Honduran Political Crisis, June 2009–January 2010* (Washington, D.C.: Library of Congress, 2010).
31. See, for instance, this article: "The Comandante's Commandments," *Economist*, November 9, 2013, https://www.economist.com/the-americas/2013/11/09/the-comandantes-commandments.
32. Lizarralde, "Organisational System and Performance."

33. Charvériat and Inter-American Development Bank, *Natural Disasters*.
34. Centre for Research on the Epidemiology of Disasters—CRED and The UN Office for Disaster Risk Reduction—UNISDR, *Economic Losses, Poverty and Disasters, 1998–2017* (New York: UN Office for Disaster Risk Reduction and Prevention, 2018), https://www.preventionweb.net/files/61119_credeconomiclosses.pdf.
35. GDP sometimes increases a few years after the disaster (notably due to more investment), but several other economic factors eventually undermine the economy in the long run.
36. Charvériat and Inter-American Development Bank, *Natural Disasters*.
37. See a report on this subject here: Rachel Cox, "New Data Reveals 197 Land and Environmental Defenders Murdered in 2017," Global Witness, February 2, 2018, https://www.globalwitness.org/en/blog/new-data-reveals-197-land-and-environmental-defenders-murdered-2017/.
38. See, for instance, this article: "Piden al Gobierno firmar acuerdo para proteger a líderes ambientales," Semana Sostenible, *Semana*, August 21, 2019, https://sostenibilidad.semana.com/medio-ambiente/articulo/que-es-el-acuerdo-de-escazu-y-por-que-colombia-debe-firmalo/45437.
39. Edith Champagne and Houssam Hariri, "Storm Flooding Brings Misery to Syrian Refugees in Lebanon," UNHCR, January 11, 2019, https://www.unhcr.org/news/latest/2019/1/5c386d6d4/storm-flooding-brings-misery-syrian-refugees-lebanon.html.
40. This is what many experts in the field call the "forensics" of disasters.
41. Find the updated figure here: "Cuba: GDP per Capita (Constant 2000 US$)," Trading Economics, https://tradingeconomics.com/cuba/gdp-per-capita-constant-2005-us$-wb-data.html
42. Martha Thompson and Izaskun Gaviria, *Weathering the Storm: Lessons in Risk Reduction from Cuba* (Boston: Oxfam, 2004); Guillermo Mesa, "The Cuban Health Sector & Disaster Mitigation," *MEDICC Review* 10, no. 3 (2008): 5–8; Gonzalo Lizarralde et al., "A Systems Approach to Resilience in the Built Environment: The Case of Cuba," *Disasters* 39, no. s1 (2015): s76-s95; Izaskun Gaviria, "Lesson from Cuba: Natural Disaster Prevention, Mitigation and Recovery" (master's thesis, Brandeis University, 2003).
43. Richard Stone, "Climate Adaptation: Cuba's 100-Year Plan for Climate Change," *Science* 359, no. 6372 (2018): 144–45.
44. I have changed the names of the villages to protect the identities of people and decision makers I met in Cuba. But the information about the case is real and has been rigorously documented.
45. See, for instance, Oscar Figueredo Reinaldo and Gabriela Roig Rosell, "A qué se destinará el presupuesto del estado cubano en el 2018?," Cuba Debate, February 7, 2018, http://www.cubadebate.cu/especiales/2018/02/07/a-que-se-destinara-el-presupuesto-del-estado-cubano-en-el-2018-video-e-infografia/—.XHFWWdFCc6g.
46. After Hurricane Irma, one single town in Cuba obtained 75 percent of the 2018 provincial budget for reconstruction. In Cuba, the portion of the 2018 budget linked to the functioning of the state and subsidies that cover disaster risk reduction and management is about 20 percent. See Reinaldo and Rosell, "A qué se destinará el

presupuesto del estado cubano en el 2018?" Similar budget increases also exist in the United States. It is estimated that annual hurricane losses have grown from $5 billion in the 1940s to more than $40 billion in the 1990s. See Stanley A Changnon et al., "Human Factors Explain the Increased Losses from Weather and Climate Extremes," *Bulletin of the American Meteorological Society* 81, no. 3 (2000): 437–442.

47. This is related to a law known as Decreto Ley N° 212—Gestión de la zona costera.
48. This section is based on what scholars call the "behavioralism" approach to disaster studies, which explores how people respond to risk and destruction.
49. Ming-Chou Ho et al., "How Do Disaster Characteristics Influence Risk Perception?," *Risk Analysis: An International Journal* 28, no. 3 (2008): 635–643; Experts identify two categories of components that influence risk perception: situational factors and cognitive factors. See Graham A. Tobin, *Natural Hazards: Explanation and Integration* (New York: Guilford, 1997).
50. Paul Slovic, "Perception of Risk," *Science* 236, no. 4799 (1987): 280–285.
51. David Garland, "The Rise of Risk," in *Risk and Morality*, ed. R. Erickson (Toronto: University of Toronto Press, 2003): 48–86.
52. This section is mostly based on Tamás Vasvári, "Risk, Risk Perception, Risk Management—A Review of the Literature," *Public Finance Quarterly* 60, no. 1 (2015): 29–48.; Paul Slovic, "Perception of Risk," *Science* 236, no. 4799 (1987): 280–285; Heather Bell, *Efficient and Effective? The Hundred Year Flood in the Communication and Perception of Flood Risk* (Tampa, FL: Scholar Commons, 2004); Benjamin Wisner, "Vulnerability as Concept, Model, Metric, and Tool," in *Oxford Research Encyclopedia of Natural Hazard Science* (Oxford: Oxford University Press, 2016), https://doi.org/10.1093/acrefore/9780199389407.013.25; and Graham A Tobin, *Natural Hazards: Explanation and Integration* (New York: Guilford Press, 1997).
53. Kate Burningham et al., " 'It'll Never Happen to Me': Understanding Public Awareness of Local Flood Risk," *Disasters* 32, no. 2 (2008): 216–238; Tali Sharot, "The Optimism Bias," *Current Biology* 21, no. 23 (2011): R941–R945.
54. Roger G. Noll and James E. Krier, "Some Implications of Cognitive Psychology for Risk Regulation," *Journal of Legal Studies* 19, no. S2 (1990): 747–779.
55. See, for instance, Hanna A Ruszczyk, "A Continuum of Perceived Urban Risk—From the Gorkha Earthquake to Economic Insecurity," *Environment and Urbanization* 30, no. 1 (2018): 317–332. Also see Gonzalo Lizarralde and the Invisible Houses, "ADAPTO: Climate Change and Social Justice in Latin America," YouTube video, 1:23, February 13, 2020, https://www.youtube.com/watch?v=rv2OVxa57KE.
56. Anna Scolobig et al., "The Missing Link Between Flood Risk Awareness and Preparedness: Findings from Case Studies in an Alpine Region," *Natural Hazards* 63, no. 2 (2012): 499–520.
57. Michael Siegrist and Heinz Gutscher, "Natural Hazards and Motivation for Mitigation Behavior: People Cannot Predict the Affect Evoked by a Severe Flood," *Risk Analysis: An International Journal* 28, no. 3 (2008): 771–778.
58. Makarand R Paranjape, " 'Natural Supernaturalism?' The Tagore–Gandhi Debate on the Bihar Earthquake," *Journal of Hindu Studies* 4, no. 2 (2011): 176–204.

59. Ted Steinberg, *Acts of God: The Unnatural History of Natural Disaster in America* (Oxford: Oxford University Press, 2006).
60. Find a pertinent article here: Heather Saul, "Susanne Atanus, Who Believes Gay Rights Cause Tornadoes and Autism Is a Punishment from God, Wins Illinois Primary," *Independent* (UK), March 20, 2014, https://www.independent.co.uk/news/world/americas/susanne-atanus-who-believes-gay-rights-cause-tornadoes-and-autism-is-a-punishment-from-god-wins-9204598.html.
61. See an online debate on this subject here: "Natural Disasters or 'Acts of God'?," Room for Debate, *New York Times*, November 18, 2013, https://www.nytimes.com/roomfordebate/2013/11/18/natural-disasters-or-acts-of-god.
62. Kenneth Hewitt, "The Idea of Calamity in a Technocratic Age," *Interpretations of Calamity from the Viewpoint of Human Ecology* 1 (1983): 3–32, at 6.
63. Piers M. Blaikie et al., *At Risk: Natural Hazards, People's Vulnerability, and Disasters* (New York: Routledge, 1994).
64. Omar Darío Cardona, "La necesidad de repensar de manera holística los conceptos de vulnerabilidad y riesgo," in *International Work-Conference on Vulnerability in Disaster Theory and Practice* (Netherlands: Unidad Nacional para la Gestión del Riesgo de Desastres, Presidencia de la República de Colombia, 2002).
65. Mark Pelling, *Natural Disasters and Development in a Globalizing World* (New York: Routledge, 2003), 9.
66. See, for instance, Alice Fothergill and Lori A. Peek, "Poverty and Disasters in the United States: A Review of Recent Sociological Findings," *Natural Hazards* 32, no. 1 (2004): 89–110.
67. See, for instance, Michel Masozera et al., "Distribution of Impacts of Natural Disasters Across Income Groups: A Case Study of New Orleans," *Ecological Economics* 63, nos. 2–3 (2007): 299–306.
68. See, for instance, James R Elliott and Jeremy Pais, "Race, Class, and Hurricane Katrina: Social Differences in Human Responses to Disaster," *Social Science Research* 35, no. 2 (2006): 295–321.
69. Pierre Le Hir, "Face aux colères de la nature, les peuples ne sont pas égaux," Le Monde, January 22, 2010.
70. For an analysis of social vulnerabilities in informal settlements in Central America see: Juan Pablo Pérez Sáinz et al., "Social Exclusion, Violences, and Urban Marginalisation in Central America," in *Reducing Urban Violence in the Global South*, ed. Jennifer Erin Salahub et al. (New York: Taylor and Francis, 2020), 135–154.
71. International Fund for Agricultural Development, "Hurricane Mitch: Its Effects on the Poor," Reliefweb, February 17, 1999, https://reliefweb.int/report/el-salvador/hurricane-mitch-its-effects-poor.
72. Charvériat and Inter-American Development Bank, *Natural Disasters*.
73. International Labour Organization (ILO), *Indigenous Peoples and Climate Change-from Victims to Change Agents Through Decent Work* (Geneva: ILO, 2016).
74. Timothy David Clark, ed., *Rebuilding Resilient Indigenous Communities in the RNWB: Final Report* (Cochrane, Canada: Willow Springs Strategic Solutions, 2018), http://

atcfn.ca/wp-content/uploads/2018/10/Rebuilding-Resilient-Indigenous-Communities-Executive-Summary-Final.pdf.

75. Robert Muir-Wood, *The Cure for Catastrophe: How We Can Stop Manufacturing Natural Disasters* (Philadelphia: Basic, 2015).
76. See, for instance, Richard A. Oppel Jr. et al., "The Fullest Look Yet at the Racial Inequality of Coronavirus," *New York Times*, July 5, 2020, https://www.nytimes.com/interactive/2020/07/05/us/coronavirus-latinos-african-americans-cdc-data.html.
77. Journalists reported this chain of blames in a series of articles in the *Houston Chronicle*. See David Hunn, Ryan Maye Handy, and James Osborne, "Build, Flood, Rebuild: Flood Insurance's Expensive Cycle," *Houston Chronicle*, December 9, 2016, https://www.houstonchronicle.com/news/houston-texas/houston/article/Build-flood-rebuild-flood-insurance-s-12413056.php.
78. Tatiana M. Davidson et al., "Disaster Impact Across Cultural Groups: Comparison of Whites, African Americans, and Latinos," *American Journal of Community Psychology* 52, nos. 1–2 (2013): 97–105.
79. The same principle applies to other measures such as MtCO2. See "Fossil Fuels Emissions," Global Carbon Atlas, http://www.globalcarbonatlas.org/en/CO2-emissions.
80. Rhett A. Butler, "Calculating Deforestation Figures for the Amazon," Mongabay, April 24, 2018, updated January 4, 2020, https://rainforests.mongabay.com/amazon/deforestation_calculations.html.
81. Find additional information here: "Syrian Regional Refugee Response," Operationals Portal, UNHCR, https://data2.unhcr.org/en/situations/syria/location/71.
82. See, for instance: "En cifras: Todo lo que debe saber sobre la migración venezolana," *El Tiempo* (Bogotá), November 28, 2018, https://www.eltiempo.com/mundo/venezuela/cifras-de-la-migracion-venezolana-en-colombia-septiembre-de-2018-290680.
83. Amr S Elnashai et al., "The Maule (Chile) Earthquake of February 27, 2010: Consequence Assessment and Case Studies," Mid-America Earthquake (MAE) Center, Research Report 10-04, Department of Civil and Environmental Engineering, University of Illinois at Urbana-Champaign, December 31, 2010, https://www.ideals.illinois.edu/handle/2142/18212.
84. Kevin A. Gould, M. Magdalena Garcia, and Jacob A. C. Remes, "Beyond 'Natural-Disasters-Are-Not-Natural': The Work of State and Nature After the 2010 Earthquake in Chile," *Journal of Political Ecology* 23, no. 1 (2016): 94.
85. See, for instance: Food and Agriculture Organization of the United Nations, *Lineamientos y recomendaciones para la implementación del Marco de Sendai para la Reducción del Riesgo de Desastres en el Sector Agrícola y Seguridad Alimentaria y Nutricional América Latina y el Caribe* (Santiago, Chile: Food and Agriculture Organization of the United Nations, 2017).
86. Pia Rinne and Anja Nygren, "From Resistance to Resilience: Media Discourses on Urban Flood Governance in Mexico," *Journal of Environmental Policy & Planning* 18, no. 1 (2016): 4–26, 22.
87. More information here: "Floods and Recurrence Intervals," USGS, https://water.usgs.gov/edu/100yearflood-basic.html.

88. Find the article at: Roberto Rocha, "What Are Flood Maps, and Why Are They Important?," CBC News, May 12, 2017, https://www.cbc.ca/news/canada/montreal/flood-maps-montreal-1.4113148.
89. See, for instance, Department of Regional Development and Environment Executive Secretariat for Economic and Social Affairs Organization of American States, *Primer on Natural Hazard Management in Integrated Regional Development Planning* (Washington, D.C.: Office of Foreign Disaster Assistance United States Agency for International Development, 1991).
90. Robert R. Holmes Jr. and Karen Dinicola, "100-Year Flood—It's All About Chance," USGS, April 2010, https://pubs.usgs.gov/gip/106/pdf/100-year-flood-handout-042610.pdf.
91. The hundred-year floodplain tool has been criticized on many other fronts as well. There are significant flaws in the maps and data associated with this tool. Information is sometimes insufficient or inaccurate, small-scale factors (such as street flooding or risks related to drains) are often ignored, and maps quickly become obsolete. In fact, it is estimated that in the United States, 50–66 percent of all flood losses already occur outside flood-designated areas.
92. "This number is derived using probability theory. First, [experts] calculate the probability of there not being a flood over a 30-year period. Since for each year, there is a 99 percent chance of there not being a flood, the chance that there is no flood over 30 years is 74 percent (or .99^30). The probability of a house in a 100-year floodplain being inundated at least once, then, is just the complement, so 26 percent." This analysis is provided by Maggie Koerth-Baker in "It's Time to Ditch the Concept of '100-Year Floods,'" FiveThirtyEight, August 30, 2017, https://fivethirtyeight.com/features/its-time-to-ditch-the-concept-of-100-year-floods/.
93. See report at "41 Million Americans Live in Flood Zones—Three Times the FEMA Estimate, Finds New Study," *Yale Environment 360*, March 5, 2018, https://e360.yale.edu/digest/41-million-americans-live-in-flood-zones-three-times-the-fema-estimate-finds-new-study.
94. Heather Bell, *Efficient and Effective? The Hundred Year Flood in the Communication and Perception of Flood Risk* (Tampa, FL: Scholar Commons, 2004).
95. An anecdote in Gabriel García Márquez's memoirs, *Living to Tell the Tale* (New York: Vintage, 2004).
96. See article here: "The South Asian Monsoon, Past, Present, and Future," *Economist*, June 27, 2019, https://www.economist.com/essay/2019/06/27/the-south-asian-monsoon-past-present-and-future.
97. See, for instance, "India Floods: At Least 95 Killed, Hundreds of Thousands Evacuated," BBC, August 10, 2019, https://www.bbc.com/news/world-asia-india-49306246. See also, Navin Singh Khadka, "Monsoon Season: The River Politics Behind South Asia's Floods," BBC, July 15, 2019, https://www.bbc.com/news/world-asia-india-48986799.
98. Elaine Enarson and Betty Hearn Morrow, "Why Gender? Why Women? An Introduction to Women and Disaster," in *The Gendered Terrain of Disaster: Through Women's Eyes*, ed. Elaine Enarson and Betty Hearn Morrow (London: Praeger, 1998), 1–8;

Elaine Enarson and P. G. Dhar Chakrabarti, *Women, Gender and Disaster: Global Issues and Initiatives* (New Delhi: Sage Publications India, 2009); Elaine Pitt Enarson, *Women Confronting Natural Disaster: From Vulnerability to Resilience* (Boulder, CO: Lynne Rienner, 2012).

99. Alicia Sliwinsky, "The Politics of Participation: Involving Communities in Post-Disaster Reconstruction," in *Rebuilding After Disasters: From Emergency to Sustainability*, ed. Gonzalo Lizarralde et al. (London: Taylor and Francis, 2010), 188–207.
100. Jean-Germain Gros, "Haiti: The Political Economy and Sociology of Decay and Renewal," *Latin American Research Review* 35, no. 3 (2000): 211–226.
101. Gregory Bankoff, "Rendering the World Unsafe: 'Vulnerability' as Western Discourse," *Disasters* 25, no. 1 (2001): 19–35, 25.
102. Gonzalo Lizarralde et al., eds., *Rebuilding After Disasters: From Emergency to Sustainability* (London: Taylor & Francis, 2009), 286.
103. Amartya Sen, *Development as Freedom* (New York: Anchor, 1999); Amartya Sen, *The Idea of Justice* (Cambridge, MA: Belknap Press, 2009).
104. In Sen's definition, capabilities are attainable outcomes, which are the joint implications of environmental opportunities and a person's abilities.
105. Gonzalo Lizarralde and Michel-Max Raynaud, "The Capability Approach in Housing Development and Reconstruction," in *Post Earthquake Reconstruction: Lessons Learnt and Way Forward* (Ahmedabad, India: Government of Gujarat, 2011); M. Nussbaum, "Human Rights and Human Capabilities," *Harvard Human Rights Journal* 20 (2007): 21–22.
106. Bankoff, "Rendering the World Unsafe," 19
107. Pelling, *Natural Disasters.*
108. Ana Gabriela Fernández et al., "Comunidad, vulnerabilidad y reproducción en condiciones de desastre: Abordajes desde América Latina y el Caribe," *Íconos. Revista de ciencias sociales*, no. 66 (2020): 7–29.
109. Brené Brown, "The Power of Vulnerability: Teachings of Authenticity, Connection, and Courage," filmed June 2010 at TEDxHouston, Ted video, 20:04, https://www.ted.com/talks/brene_brown_the_power_of_vulnerability?language=en.
110. Jason von Meding and Heidi Harmon, "Who's Afraid of Vulnerability? Reframing Vulnerability as a Strength Is What Makes Transformation Possible," Open Democracy, June 7, 2020, https://www.opendemocracy.net/en/transformation/whos-afraid-of-vulnerability/.
111. See, for instance, Donna Jeanne Haraway, *Primate Visions: Gender, Race, and Nature in the World of Modern Science* (New York: Routledge, 1989).
112. Paris Hilton (@ParisHilton), "This wild fire in LA is terrifying! My house is now being evacuated to get all my pets out of there safely. Thank you to all the firefighters who are risking their lives to save ours. You are true heroes!," Twitter, December 6, 2017, 1:59 p.m., https://twitter.com/ParisHilton/status/938482971908198400.
113. See, for instance, Jack Nicas and Thomas Fuller, "Wildfires Become Deadliest in California History," *New York Times*, November 12, 2018, https://www.nytimes.com/2018/11/12/us/california-fires-camp-fire.html.

114. Find the article here: "California's Wildfires and the New Abnormal," *Economist*, November 17, 2018, https://www.economist.com/united-states/2018/11/17/californias-wildfires-and-the-new-abnormal.
115. Pelling, *Natural Disasters*, 9
116. Paula Dunbar et al., "2011 Tohoku Earthquake and Tsunami Data Available from the National Oceanic and Atmospheric Administration/National Geophysical Data Center," *Geomatics, Natural Hazards and Risk* 2, no. 4 (2011): 305–323.
117. See the article here: Niraj Chokshi, "Neil Young and Miley Cyrus Among Celebrities Who Lost Homes in California Wildfires," *New York Times*, November 13, 2018, https://www.nytimes.com/2018/11/13/us/celebrities-lost-homes-california-fires.html.
118. See an article about this subject here: "California's Wildfires and the New Abnormal," *Economist*, November 17, 2018, https://www.economist.com/united-states/2018/11/17/californias-wildfires-and-the-new-abnormal.
119. Anthony L Westerling et al., "Warming and Earlier Spring Increase Western US Forest Wildfire Activity," *Science* 313, no. 5789 (2006): 940–943.
120. See Anthony LeRoy Westerling, "Wildfires in West Have Gotten Bigger, More Frequent, and Longer Since the 1980s," *The Conversation*, May 23, 2016, https://theconversation.com/wildfires-in-west-have-gotten-bigger-more-frequent-and-longer-since-the-1980s-42993.
121. Bill Gabbert, "At Least 50 People Killed in Wildfires Near Athnes, Greece," Wildfire Today, July 24, 2018, https://wildfiretoday.com/2018/07/24/at-least-50-people-killed-in-wildfires-near-athens-greece/.
122. "Costas," Greenpeace, https://es.greenpeace.org/es/trabajamos-en/oceanos/costas/
123. Blaikie, *At Risk*.
124. Blaikie, *At Risk*, 50.
125. Mihir Zaveri and Emily S. Rueb, "How Many Animals Have Died in Australia's Wildfires?," *New York Times*, January 11, 2020, https://www.nytimes.com/2020/01/11/world/australia/fires-animals.html.
126. Some previous disaster scholars have tried, of course, to connect environmental degradation and disasters. See, for instance, Maria Augusta Fernández and La Red, eds., *Ciudades en riesgo: Degradación ambiental, riesgos urbanos y desastres* (Quito, Ecuador: La Red and USAID, 1996).

2. Change: "They Want to Build Something Modern Here"

1. The stories of victims and survivors of Lac-Mégantic's tragedy were beautifully portrayed by journalists of the *Globe and Mail*. See Justin Giovannetti, "Last Moments of Lac-Mégantic: Survivors Share Their Stories," *Globe and Mail* (Canada), November 28, 2013, https://www.theglobeandmail.com/news/national/lac-megantic-musi-cafe/article15656116/.
2. Jenna L. Currie-Mueller, "Pointing Fingers Across the Tracks: An Examination of Strategic Messages in the Lac-Mégantic Rail Disaster," *Journal of Risk Research* 21, no. 10 (2018): 1197–1216.

3. Mélissa Généreux et al., "The Public Health Response During and After the Lac-Mégantic Train Derailment Tragedy: A Case Study," *Disaster Health* 2, nos. 3–4 (2014): 113–120; Jean-Paul Lacoursière et al., "Lac-Mégantic Accident: What We Learned," *Process Safety Progress* 34, no. 1 (2015): 2–15.
4. Galvez-Cloutier Rosa et al., "Lac-Mégantic: Analyse De L'urgence Environnementale, Bilan Et Évaluation Des Impacts," *Canadian Journal of Civil Engineering* 41, no. 6 (2014): 531–539.
5. Geneviève Brisson and Emmanuelle Bouchard-Bastien, "Post-Catastrophe Perceptions of Risk and Development in Lac-Mégantic, Québec," in *ExtrACTION: Impacts, Engagements, and Alternative Futures*, ed. Kirk Jalbert, Anna Willow, David Casagrande, and Stephanie Paladino (New York: Routledge, 2017), 123–136.
6. Alain Deneault and Aaron Barcant, "Tu n'as rien vu à Lac-Mégantic," *Liberté* 303 (Spring edition, 2014): 47–48, 48.
7. Gilbert Liette, "The Crisis After the Crisis: Neoliberalized Discourses of Urgency, Risk and Resilience in the Reconstruction of Lac-Mégantic," *Revue générale de droit* 48 (2018): 155–175, 155.
8. Anne-Julie Tremblay, "Le vécu des personnes âgées de 65 ans ou plus ayant été exposées à la tragédie ferroviaire de Lac-Mégantic" (master's thesis, Université du Québec à Chicoutimi, 2019).
9. M. Genereux et al., "Two Years after the Train Derailment: Lac-Megantic Residents Are Still Suffering," *European Journal of Public Health* 26, no. 1 (2016): ckw170.005.
10. Pierre Thibault and Ville de Lac-Mégantic, *Annexe 2—Objectifs Et Critères P.I.I.A.—010—Secteur Résidentiel Du Centre-Ville* (Lac-Mégantic, Canada: Ville de Lac-Mégantic, 2015).
11. Jennifer E. Duyne Barenstein and Esther Leemann, *Post-Disaster Reconstruction and Change: Communities' Perspectives* (London: CRC, 2012); Jennifer Duyne Barenstein et al., *Safer Homes, Stronger Communities: A Handbook for Reconstruction after Natural Disasters* (Washington, D.C.: World Bank, 2010); Jennifer E. Duyne Barenstein, "Housing Reconstruction in Post-Earthquake Gujarat: A Comparative Analysis," *Humanitarian Practice Network Paper* 54 (March 2006): 1–36; Aparna Tandon, "Post-Disaster Damage Assessment of Cultural Heritage: Are We Prepared?," in *18th Triennial Conference* (Rome: ICOM-CC, 2017), https://www.iccrom.org/sites/default/files/2017-12/tandon_2017_post-disaster_damage_assessment_icomcc_2017.pdf; Yu Wang, "A Sustainable Approach for Post-Disaster Rehabitation of Rural Heritage Settlements," *Sustainable Development* 24, no. 5 (2016): 319–329.
12. Krzysztof Kaniasty and Fran Norris, "The Experience of Disaster: Individuals and Communities Sharing Trauma," in *Response to Disaster: Psychosocial, Community, and Ecological Approaches*, ed. Bernard Lubin and Richard Gist (New York: Routledge, 1999), 25–61; Barbara B. Brown and Douglas D. Perkins, "Disruptions in Place Attachment," in *Place Attachment*, ed. Irwin Altman and Setha Low (New York: Plenum, 1992), 279–304.
13. Naomi Klein, *The Shock Doctrine: The Rise of Disaster Capitalism* (New York: Macmillan, 2007).

14. See, for instance, their Facebook group: Le Carré Bleu Mégantic, https://www.facebook.com/lecarrebleulacmegantic/.
15. Suzanne Wilkinson et al., "Reconstruction Following Earthquake Disasters," in *Encyclopedia of Earthquake Engineering*, ed. Michael Beer et al. (New York: Springer, 2015), 1–11; I. Thomas Maret and T. Cadoul, "Résilience Et Reconstruction Durable : Que Nous Apprend La Nouvelle-Orléans?," *Annales de Géographie* 663 (2008): 104–124; Gonzalo Lizarralde, F. Kikano, M. Fayazi, I. Thomas, "Meta Patterns in Post-Disaster Housing Reconstruction and Recovery," in *Post-Disaster Housing*, ed. lka Sapat and Ann-Margaret Esnard (Miami: CRC, 2016), 229–242; Gonzalo Lizarralde, "Decentralizing (Re)Construcion: Agriculture Cooperatives as a Vehicle for Reconstruction in Colombia," in *Building Back Better: Delivering People-Centered Housing Reconstruction at Scale*, ed. Michal Lyons and Theo Schilderman (London: Practical Action, 2010), 191–214; David Alexander et al., eds., "Post-Disaster Reconstruction: Meeting Stakeholder Interests" (proceedings of 3rd International I-Rec Conference, Scuola di Sanità Militare, Florence, Italy, May 17–19, 2006).
16. Ilan Kelman et al., eds., *The Routledge Handbook of Disaster Risk Reduction Including Climate Change Adaptation* (London: Routledge, 2017); Rohit Jigyasu, "Appropriate Technology for Post-Disaster Reconstruction," in *Rebuilding after Disasters: From Emergency to Sustainability*, ed. Gonzalo Lizarralde et al. (New York: Taylor and Francis, 2010), 49–69; Teddy Boen and Rohit Jigyasu, "Cultural Considerations for Post Disaster Reconstruction Post-Tsunami Challenges," Paper presented at the UNDP Conference, 2005. http://www.adpc.net/IRC06/2005/4-6/TBindo1.pdf; Cut Dewi, "Rethinking Architectural Heritage Conservation in Post-Disaster Context," *International Journal of Heritage Studies* 23, no. 6 (2017): 587–600; Robert Soden and Austin Lord, "Mapping Silences, Reconfiguring Loss: Practices of Damage Assessment & Repair in Post-Earthquake Nepal," *Proceedings of the ACM on Human-Computer Interaction* 2, no. CSCW (2018): 1–21.
17. See, for example, "El fallo 'histórico' del río Bogotá," *Semana*, April 5, 2014, https://www.semana.com/nacion/articulo/limpieza-del-rio-bogota-no-esta-tan-en-panales-como-la-gente-cree/382708-3. Also see Wikipedia, s.v. "Tequendama Falls," last modified December 22, 2020, 00:42, https://en.wikipedia.org/wiki/Tequendama_Falls, note 5.
18. In 2007, the Associación Nacional de Instituciones Financieras (ANIF) reported that housing prices in Bogotá had increased about 25 percent between 2012 and 2016, one of the highest increases in the country, higher than the national average and the increases in Cali and Medellín. See Sergio Clavijo, ed., with Ekaterina Cuéllar and Daniel Beltrán, "Los premios de la vivienda en Colombia (2005–2017)," *RASEC: Reporte Anif sector construcción*, December 2017, http://anif.co/sites/default/files/publicaciones/private/restricted/2017/12/rasec191.pdf.
19. Acevedo Bohórquez Acevedo, *Territorio y sociedad: El caso del POT de la Ciudad de Bogotá* (Bogotá, Colombia: Universidad Nacional de Colombia, 2003).
20. See a report on the rivers's condition here: "Renace el Río Bogotá: Un esfuerzo de todos," https://especiales.semana.com/rio_bogota/.

21. See, for instance, Observatorio de Conflictos Ambientales and Instituto de Estudios Ambientales, "Los peligros de modificar la Reserva Thomas van der Hammen," Semana Sostenible, *Semana*, June 26, 2018, https://sostenibilidad.semana.com/opinion/articulo/los-peligros-de-modificar-la-reserva-thomas-van-der-hammen/41085.
22. Meteorología y Estudios Ambientales IDEAM-Instituto de Hidrología, *Memoria Descriptiva—Mapas De Inundación Departamento De Cundinamarca* (Bogotá, Colombia: Minambiente, 2013).
23. See an analysis here: Daniel Bernal, "Humedales Bogotá," Fundación Humedales Bogotá, December 13, 2011, http://humedalesbogota.com/2011/12/13/porque-nos-inundan-los-rios-el-caso-de-la-universidad-de-la-sabana/.
24. " 'La van der Hammen no tiene nada distinto a cualquier otro potrero': Peñalosa," *El Espectador* (Bogotá), July 25, 2016, https://www.elespectador.com/noticias/bogota/van-der-hammen-no-tiene-nada-distinto-cualquier-otro-po-articulo-645224.
25. Ana María Cuevas, "Propuesta de Alcaldía Peñalosa hará realidad una Van der Hammen con mucha más área de reserva," Bogota.gov, April 2, 2018, http://www.bogota.gov.co/temas-de-ciudad/planeacion/propuesta-de-alcaldia-penalosa-para-reserva-van-der-hammen.
26. "Las razones de la Alcaldía para no intervenir la Van der Hammen," *El Tiempo* (Bogotá), January 15, 2020, https://www.eltiempo.com/bogota/alcaldesa-claudia-lopez-no-intervendra-reserva-thomas-van-der-hammen-452316.
27. Michel Foucault, "Le Sujet Et Le Pouvoir," *Dits et écrits* 4 (1994): 222–242.
28. James Scott, *Seeing Like a State: How Certain Schemes to Improve the Human Condition Have Failed* (New Haven, CT: Yale University Press, 1998), 78, 77.
29. Gonzalo Lizarralde, *The Invisible Houses: Rethinking and Designing Low-Cost Housing in Developing Countries* (London: Routledge, 2014).
30. Jason Hickel, *The Divide: A Brief Guide to Global Inequality and Its Solutions* (London: Random House, 2017).
31. Liette, "The Crisis After the Crisis"; Daniel Sage et al., "Securing and Scaling Resilient Futures: Neoliberalization, Infrastructure, and Topologies of Power," *Environment and Planning D: Society and Space* 33 (2015): 494–511; John Gledhill, "Neoliberalism," in *A Companion to the Anthropology of Politics*, ed. David Nugent and Joan Vincent (Victoria, Canada: Blackwell, 2004), 332–348; Carlos Arze and Tom Kruse, "The Consequences of Neoliberal Reform," *NACLA Report on the Americas* 38, no. 3 (2004): 23–28.
32. Hickel, *The Divide*, 141.
33. Regulation experts seem to particularly enjoy identifying the very first construction code. According to some accounts, the first construction code appeared in Rome after the Great Fire of 64 AD. But I suspect other more ancient traces will be found by enthusiastic code historians.
34. David Drengenberg and Gene Corley, "Evolution of Building Code Requirements in a Post 9/11 World," *CTBUH Journal*, no. 3 (2011): 32–35.
35. William Solesbury, *Policy in Urban Planning: Structure Plans, Programmes and Local Plans* (Oxford: Pergamon Press, 1974).

36. Marisa O. Ensor, *The Legacy of Hurricane Mitch: Lessons from Post-Disaster Reconstruction in Honduras* (Tucson: University of Arizona Press, 2009).
37. CECI, *Projet De Reconstruction De Maisons Au Profit Des Sinistrés De L'ouragan Mitch: Rapport Narratif Final* (Montreal: CECI, 2001).
38. Gonzalo Lizarralde, "Organisational System and Performance of Post-Disaster Reconstruction Projects" (PhD thesis, Université de Montréal, 2004), 460.
39. Lizarralde, *Invisible Houses*.
40. The case of abandoned water wells in Africa is clearly portrayed in Ken Stern, *With Charity for All: Why Charities Are Failing and a Better Way to Give* (New York: Anchor, 2013).
41. Gro Harlem Brundtland, *Report of the World Commission on Environment and Development: "Our Common Future"* (New York: United Nations, 1987).
42. Ken Stern, *With Charity for All: Why Charities Are Failing and a Better Way to Give* (New York: Anchor, 2013).
43. See a pertinent article here: Amanda Kelly, "Quebec Red Cross Receives $7.2 Million in Donations for Lac-Megantic," *Global News* (Canada), July 25, 2013, https://globalnews.ca/news/740987/quebec-red-cross-receives-7-2-million-in-donations-for-lac-megantic/.
44. John Telford et al., *Learning Lessons from Disaster Recovery: The Case of Honduras* (Washington D.C.: World Bank, 2004).
45. Adrian Wood et al., *Evaluating International Humanitarian Action* (New York: Zed, 2001).
46. Figures change according to different sources. Here I have relied on "Net Official Development Assistance and Official Aid Received (Current US$)," World Bank, https://data.worldbank.org/indicator/DT.ODA.ALLD.CD?end=2017&start=1987.
47. Jeffrey Sachs, *The End of Poverty: How We Can Make It Happen in Our Lifetime* (New York: Penguin, 2005).
48. Nicholas Kristof, "How Can We Help the World's Poor," *New York Times*, November 20, 2009, https://www.nytimes.com/2009/11/22/books/review/Kristof-t.html.
49. Sachs, *The End of Poverty*, 367.
50. Steven Pinker, *Enlightenment Now: The Case for Reason, Science, Humanism, and Progress* (New York: Viking, 2018), 124.
51. See Hannah Ritchie, "Global Inequalities in CO_2 Emissions," Our World in Data, October 16, 2018, https://ourworldindata.org/co2-by-income-region.
52. For a graphic illustration of this, see "China Is Surprisingly Carbon-Efficient—But Still the World's Biggest Emitter," *Economist*, May 25 2019, https://www.economist.com/graphic-detail/2019/05/25/china-is-surprisingly-carbon-efficient-but-still-the-worlds-biggest-emitter.
53. William Easterly, *The White Man's Burden: Why the West's Efforts to Aid the Rest Have Done So Much Ill and So Little Good* (New York: Penguin, 2006).; Dambisa Moyo, *Dead Aid: Why Aid Makes Things Worse and How There Is Another Way for Africa* (New York: Penguin, 2010).
54. Easterly, *The White Man's Burden*, 12.

55. Gonzalo Lizarralde, "Does Aid (Actually) Aid in Avoiding Disasters and Rebuilding After Them?," OD Online Debates, 2019, https://oddebates.com/8th-debate/.
56. Gregory Bankoff, "Rendering the World Unsafe: 'Vulnerability' as Western Discourse," *Disasters* 25, no. 1 (2001): 19–35, 27.
57. Anthony J. Onwuegbuzie and Nancy L. Leech, "On Becoming a Pragmatic Researcher: The Importance of Combining Quantitative and Qualitative Research Methodologies," *International Journal of Social Research Methodology* 8, no. 5 (2005): 375–387.
58. Sharachchandra M Lélé, "Sustainable Development: A Critical Review," *World Development* 19, no. 6 (1991): 607–621.
59. Jakob Svensson, "Aid, Growth and Democracy," *Economics & Politics* 11, no. 3 (1999): 275–297.
60. Abhijit Vinayak Banerjee and Esther Duflo, *Poor Economics: A Radical Rethinking of the Way to Fight Global Poverty* (New York: PublicAffairs, 2011).
61. Paul Collier and David Dollar, "Aid Allocation and Poverty Reduction," *European Economic Review* 46, no. 8 (2002): 1475–1500.
62. David Collier Dollar and Paul Collier "Can the World Cut Poverty in Half? How Policy Reform and Effective Aid Can Meet International Development Goals," (Policy Research working paper no. 2403, World Bank, Washington, D.C., 2000), https://openknowledge.worldbank.org/handle/10986/19823.
63. Banerjee and Duflo, *Poor Economics*.
64. See, for instance: Robert Calderisi, "Why Foreign Aid and Africa Don't Mix," CNN, August 18, 2010, http://www.cnn.com/2010/OPINION/08/12/africa.aid.calderisi/index.html.
65. Hickel, *The Divide*.
66. Deborah Brautigam, "China, Africa and the International Aid Architecture" (African Development Bank Group working paper 107, 2010).
67. Mahmood Fayazi, "Household Recovery and Housing Reconstruction After the 2003 Bam Earthquake in Iran" (PhD thesis, Université de Montréal, 2018).
68. Sebastian Horn et al., *China's Overseas Lending* (Cambridge, MA: National Bureau of Economic Reseach, 2019).
69. See "Chinese Loans to Poor Countries Are Surging," *Economist*, July 12, 2019, https://www.economist.com/graphic-detail/2019/07/12/chinese-loans-to-poor-countries-are-surging.
70. Stern, *With Charity for All*.
71. More about this can be found here: "History," IFRC.org, http://www.ifrc.org/en/who-we-are/history/.
72. Lizarralde, "Meta Patterns"; Cassidy Johnson et al., "A Systems View of Temporary Housing Projects in Post-Disaster Reconstruction," *Construction Management and Economics* 24, no. 4 (2006): 367–378; Cassidy Johnson and Gonzalo Lizarralde, "Post-Disaster Housing and Reconstruction," *International Encyclopedia of Housing and Home*, no. 46 (2012): 340–346; Cassidy Johnson, "Impacts of Prefabricated Temporary Housing after Disasters: 1999 Earthquakes in Turkey," *Habitat International* 31, no. 1 (2007): 36–52.

73. Mahmood Fayazi and Gonzalo Lizarralde, "Conflicts Between Recovery Objectives: The Case of Housing Reconstruction After the 2003 Earthquake in Bam, Iran," *International Journal of Disaster Risk Reduction* 27 (2018): 317–328; Gonzalo Lizarralde et al., eds. *Rebuilding After Disasters: From Emergency to Sustainability* (London: Taylor & Francis, 2009) 286.
74. Johnson et al., "A Systems View of Temporary Housing Projects"; Johnson, "Impacts of Prefabricated Temporary Housing after Disasters."
75. I have modified the name of the organization and its founder to protect its employers from unnecessary criticism.
76. Kate Stohr and Cameron Sinclair, *Design Like You Give a Damn* (New York: Metropolis, US, 2006).
77. The same principle applies, to a degree, to other forms of expression that require creativity, such as journalism or marketing. An individual who despizes sensationalist journalism can most often avoid it or at least refrain from consuming it.
78. For an example of authors who consider design to have its own value, see: Philippe D'Anjou, *Design Ethics Beyond Duty and Virtue* (Newcastle Upon Tyne, UK: Cambridge Scholars Publishing, 2017).

3. Sustainability: "They Often Come Here With Their Talk About Green Solutions"

1. See the video here: Tangram 3DS, "Harvest City, Haiti 'A Concept to Recovery,'" YouTube video, 6:40, December 20, 2010, https://www.youtube.com/watch?v=srwFzw8206w.
2. Alexei Barrionuevo, "After Deadly Mudslides in Brazil, Concern Turns to Preparedness," *New York Times*, January 16, 2011, https://www.nytimes.com/2011/01/17/world/americas/17brazil.html.
3. Tom Phillips, "Brazil Landslides Leave Hundreds of People Dead," *Guardian*, January 12, 2011, https://www.theguardian.com/world/2011/jan/12/brazil-landslide-leaves-115-dead.
4. "Deforestation in Brazil," Belgian Earth Observation, 2009, https://eo.belspo.be/en/deforestation-brazil.
5. Herton Escobar, "Brazil's Deforestation Is Exploding—and 2020 Will Be Worse," *Science*, November 22, 2019, https://www.sciencemag.org/news/2019/11/brazil-s-deforestation-exploding-and-2020-will-be-worse.
6. D. M. Silva Matos et al., "Fire and Restoration of the Largest Urban Forest of the World in Rio De Janeiro City, Brazil," *Urban Ecosystems* 6, no. 3 (2002): 151–161.
7. "Rio de Janeiro Sees Growth of Favela Communities," *Rio Times*, May 11, 2015, https://riotimesonline.com/brazil-news/rio-politics/rio-de-janeiro-sees-growth-of-favela-communities/.
8. Shasta Darlington, "Eco-Wall or Segregation: Rio Plan Stirs Debate," CNN, December 9, 2009, http://www.cnn.com/2009/WORLD/americas/12/09/brazil.ecowall/index.html.
9. Gro Harlem Brundtland, *Report of the World Commission on Environment and Development: "Our Common Future"* (New York: United Nations, 1987). Hereafter referred to as Brundtland Report.

10. Brundtland Report, 259.
11. Sheila Bonini, *The Business of Sustainability* (New York: McKinsey & Company, 2011), https://www.mckinsey.com/business-functions/sustainability/our-insights/the-business-of-sustainability-mckinsey-global-survey-results#.
12. See the graph at: U.S. Green Building Council, "Cumulative Number of LEED Registrations in the U.S. from 2000 to 2019," Statista, November 19, 2020, https://www.statista.com/statistics/323383/leed-registered-projects-in-the-united-states/.
13. See full article at: Marisa Long, "Green Building Accelerates Around the World, Poised for Strong Growth by 2021," U.S. Green Building Council, November 13, 2018, https://www.usgbc.org/articles/green-building-accelerates-around-world-poised-strong-growth-2021.
14. Stephanie Vierra, "Green Building Standards and Certification Systems," Whole Building Design Guide, last updated August 5, 2019, https://www.wbdg.org/resources/green-building-standards-and-certification-systems.
15. Dodge Data & Analytics, *World Green Building Trends 2018* (Bedford, MA: Dodge Data & Analytics, 2018).
16. Nielsen, *The Sustainability Imperative—New Insights on Consumer Expectations* (New York: Nielsen Global Media, 2015).
17. Daniel Mahler, "An Emerging Retail Trend Is Key for Attracting Millennials," *Business Insider*, October 27, 2015, https://www.businessinsider.com/how-important-is-sustainability-to-millennials-2015-10.
18. Cone Communications, *2014 Cone Communications Food Issues Trend Tracker* (Boston: Cone Communications, 2014).
19. Cone Communications, *Cone Communications Corporate Social Responsibility Study* (Boston: Cone Communications, 2017).
20. Find the explanation here: https://www.epo.org/news-events/in-focus/sustainable-technologies/green-construction.html.
21. The declaration eventually lapsed after three months, as it was not been signed by at least half of the MEPs. See B. Boissière, "The EU Sustainable Development Discourse–An Analysis," *L'Europe en formation* no. 2 (2009): 23–39. The declaration can be found at: http://www.europarl.europa.eu/sides/getDoc.do?pubRef=-//EP//NONSGML+WDECL+P6-DCL-2008-0064+0+DOC+PDF+V0//EN&language=EN.
22. Find a related article here: Rebecca Lindsey and LuAnn Dahlman, "Climate Change: Global Temperature," Climate.gov, August 14, 2020, https://www.climate.gov/news-features/understanding-climate/climate-change-global-temperature.
23. "Arctic Sea Ice Minimum," Global Climate Change, NASA, last updated December 3, 2020, https://climate.nasa.gov/vital-signs/arctic-sea-ice/.
24. "Sea Level," Global Climate Change, NASA, last updated December 3, 2020, https://climate.nasa.gov/vital-signs/sea-level/.
25. M. Grooten and R. E. Almond, eds., *Living Planet Report—2018: Aiming Higher* (Gland, Switzerland: World Wildlife Fund, 2018).
26. Greenpeace, *The Greenpeace Green Living Guide* (Toronto: Greenpeace Canada, 2007).
27. Anthony L. Andrady, "Microplastics in the Marine Environment," *Marine Pollution Bulletin* 62, no. 8 (2011): 1596–1605.

28. See figure at: Hannah Ritchie and Max Roser, "Natural Disasters," Our World in Data, revised November 2019, https://ourworldindata.org/natural-disasters.
29. Max Roser, Hannah Ritchie, and Esteban Ortiz-Ospina, "World Population Growth," Our World in Data, revised May 2019, https://ourworldindata.org/world-population-growth.
30. Melinda Harm Benson and Robin Kundis Craig, "The End of Sustainability," *Society & Natural Resources* 27, no. 7 (2014): 777–782.
31. Subhabrata Bobby Banerjee, "Who Sustains Whose Development? Sustainable Development and the Reinvention of Nature," *Organization Studies* 24, no. 1 (2003): 143–180, 144.
32. Shiv Visvanathan, "Mrs. Brundtland's Disenchanted Cosmos," *Alternatives* 16, no. 3 (1991): 377–384, 383.
33. Mike Hannis, "After Development? In Defence of Sustainability," *Global Discourse* 7, no. 1 (2017): 28–38.
34. Rupert J. Baumgartner, "Critical Perspectives of Sustainable Development Research and Practice," *Journal of Cleaner Production* 19, no. 8 (2011): 783–786; Lucas Seghezzo, "The Five Dimensions of Sustainability," *Environmental Politics* 18, no. 4 (2009): 539–556.
35. Markku Lehtonen, "Mainstreaming Sustainable Development in the OECD Through Indicators and Peer Reviews," *Sustainable Development* 16, no. 4 (2008): 241–250.
36. N. Mohan Das Gandhi et al., "Unsustainable Development to Sustainable Development: A Conceptual Model," *Management of Environmental Quality International Journal* 17, no. 6 (2006): 654–672.
37. David Griggs et al., "Policy: Sustainable Development Goals for People and Planet," *Nature* 495, no. 7441 (2013): 305.
38. C. S. Holling, "Investing in Research for Sustainability," *Ecological Applications* 3, no. 4 (1993): 552–555.
39. Find an example here: "Sustainability and Sustainable Development: What Is Sustainability and What Is Sustainable Development?," Circular Ecology, https://circularecology.com/sustainability-and-sustainable-development.html.
40. Ernest J. Yanarella et al., "Research and Solutions: 'Green' vs. Sustainability: From Semantics to Enlightenment," *Sustainability* 2, no. 5 (2009): 296–302.
41. For a detailed explanation of this principle, see Wilfred Beckerman, " 'Sustainable Development': Is It a Useful Concept?," *Environmental Values* 3, no. 3 (1994): 191–209.
42. Figures from the Australian Bureau of Statistics. See "Average Floor Area of New Residential Dwellings," Australian Bureau of Statistics, November 11, 2013, http://www.abs.gov.au/AUSSTATS/abs@.nsf/Previousproducts/8752.0Feature Article1Jun 2013.
43. André Stephan and Robert Crawford, "Size Does Matter: Australia's Addition to Big Houses Is Blowing the Energy Budget," *The Conversation*, Decxember 13, 2016, https://theconversation.com/size-does-matter-australias-addiction-to-big-houses-is-blowing-the-energy-budget-70271.
44. U.S. Department of Housing and Urban Development and U.S. Department of Commerce, *2016 Characteristics of New Housing* (Washington, D.C.:U.S. Department of Commerce, 2016), https://www.census.gov/construction/chars/pdf/c25ann2016.pdf.

45. Johanne Whitmore and Pierre-Olivier Pineau, *État De L'énergie Au Québec* (Montreal: Chaire de gestion du secteur de l'énergie, HEC, 2018).
46. International Energy Agency, *The Future of Cooling: Opportunities for Energy-Efficient Air Conditioning* (Paris :OECD/IEA, 2018), https://webstore.iea.org/the-future-of-cooling.
47. Felipe Munoz, "Global SUV Boom Continues in 2018, but Growth Moderates," Jato, February 20, 2019, https://www.jato.com/global-suv-boom-continues-in-2018-but-growth-moderates.
48. Hiroko Tabuchi, "The World Is Embracing S.U.V.s. That's Bad News for the Climate," *New York Times*, March 3, 2018, https://www.nytimes.com/2018/03/03/climate/suv-sales-global-climate.html.
49. IATA, *Iata Annual Review* (IATA, 2018), https://www.iata.org/contentassets/c81222d96c9a4e0bb4ff6ced0126f0bb/iata-annual-review-2018.pdf.
50. "Per Capita Consumption of Bottled Water in the United States from 1999 to 2019 (in Gallons)," Statista, November 26, 2020, https://www.statista.com/statistics/183377/per-capita-consumption-of-bottled-water-in-the-us-since-1999/.
51. "Municipal Waste Generation," Conference Board of Canada," data from January 2013, https://www.conferenceboard.ca/hcp/Details/Environment/municipal-waste-generation.aspx.
52. Michael J. Turner and James W. Hesford, "The Impact of Renovation Capital Expenditure on Hotel Property Performance," *Cornell Hospitality Quarterly* 60, no. 1 (2018): 25–39.
53. "Total Sales of Home Improvement in the United States from 2008 to 2023," Statista, November 24, 2020, https://www.statista.com/statistics/239753/total-sales-of-home-improvement-retailers-in-the-us/.
54. "Average Annual Home Improvement Spend Per Homeowner in the United States from 2014 to 2018 (in U.S. Dollars)," Statista, November 24, 2020, https://www.statista.com/statistics/809304/average-annual-home-improvement-spend-us/.
55. See the comment by Roland Geyer, author of the paper "Production, Use and Fate of All Plastics Ever Made." See Leslie Hook and John Reed, "Why the World's Recycling System Stopped Working," *Financial Times*, October 25, 2018, https://www.ft.com/content/360e2524-d71a-11e8-a854-33d6f82e62f8.
56. Hook and Reed, "Why the World's Recycling System Stopped Working." See also: R. Geyer et al., "Production, Use, and Fate of All Plastics Ever Made," *Science Advances* 3, no. 7 (2017): e1700782.
57. International Energy Agency, *Global Energy & Co2 Status Report: The Latest Trends in Energy and Emissions in 2018* (Paris: International Energy Agency, 2019), https://webstore.iea.org/global-energy-co2-status-report-2018.
58. Based on figures published by The World Bank in: "CO2 Emissions (Metric Tons per Capita)," World Bank, https://data.worldbank.org/indicator/EN.ATM.CO2E.PC?end=2014&start=1987. Data generated by the Carbon Dioxide Information Analysis Center, Environmental Sciences Division, Oak Ridge National Laboratory, Tennessee, United States.

59. Lighting Planners Associates, "Ecopark Lighting Masterplan: An Ecological Lighting Environment with No Light Pollution," UrbanNext, https://urbannext.net/ecopark-lighting-masterplan/.
60. Rita Obiozo and John Smallwood, "Mega Projects and the Four Sublime—The Case of the Innovative Strategy of the Biophilic Construction Site Model: The Case Study of Ecopark Eco-City, Hanoi, Vietnam," in *ITHE 2014 (Fifth) International Conference On Engineering, Project, And Production Management* (Nelson Mandela Metropolitan University: Department of Construction Management, 2014), 194–2014.
61. "Ecopark Celebrates Regional Laudation," *Vietnam Breaking News*, May 17, 2014, https://www.vietnambreakingnews.com/2014/05/ecopark-celebrates-regional-laudation/.
62. "Ecopark Shines at National Property Awards 2018," *Vietnam Plus*, April 18, 2018, https://en.vietnamplus.vn/ecopark-shines-at-national-property-awards-2018/129720.vnp.
63. Hong T. A. Nguyen et al., "Mediation Effects of Income on Travel Mode Choice: Analysis of Short-Distance Trips Based on Path Analysis with Multiple Discrete Outcomes," *Transportation Research Record* 2664, no. 1 (2017): 23–30.
64. N. Hieu Nguyen, "Large-Scale Land Acquisition Lessons Learnt from the Ecopark Project, Vietnam," in *Forum for Urban Future in Southern Asia* (Cologne, Germany: 2012).
65. Danielle Labbé, "Media Dissent and Peri-Urban Land Struggles in Vietnam: The Case of the Văn Giang Incident," *Critical Asian Studies* 47, no. 4 (2015): 495–513.
66. Claire Provost and Matt Kennard, "Inside Hanoi's Gated Communities: Elite Enclaves Where Even the Air Is Cleaner," *Guardian*, January 21, 2016, https://www.theguardian.com/cities/2016/jan/21/inside-hanoi-gated-communities-elite-enclaves-air-cleaner.
67. "Security Forces Seize Land from Vietnam Villagers," Reuters, April 24, 2012, https://www.reuters.com/article/us-vietnam-clash/security-forces-seize-land-from-vietnam-villagers-idUSBRE83N0AV20120424.
68. Radio Free Asia, "'Hired Thugs' Fire on Farmers in Vietnam Land Dispute," Ref World-UNHCR, February 11, 2014, https://www.refworld.org/docid/532adf1514.html.
69. Rachel Vandenbrink, "A Year After Clashes, Vietnamese Land Dispute Unresolved," Radio Free Asia, April 24, 2013, https://www.rfa.org/english/news/vietnam/van-giang-04242013184638.html.
70. "EcoPark Satellite City Project, Hanoi, Vietnam," Environmental Justice Atlas, last updated June 19, 2015, https://ejatlas.org/conflict/ecopark-satellite-city-project-hanoi-vietnam.
71. H. A. Tran, "From Socialist Modernism to Market Modernism?," in *Routledge Handbook of Urbanization in Southeast Asia*, ed. Rita Padawangi (London: Routledge, 2018), 249–264.
72. Provost and Kennard, "Insidde Hanoi's Gated Communities."
73. Danielle Labbé, "Once the Land Is Gone: Land Redevelopment and Livelihood Adaptations on the Outskirts of Hanoi, Vietnam," in *Balanced Growth for an Inclusive and Equitable ASEAN Community*, ed. Mely Caballero-Anthony and Richard Barichello (Singapore: RSIS Centre for NTS Studies, 2015), 148.
74. Nguyen Quang Minh et al., "The Four Villages Surrounding the Kdtm of Van Quan," in *Bridging the Gap: Towards a Better Integration of Masterplanned New Urban Areas and Urbanised Villages*, ed. D. Labbé (Vietnam: Gioi Publishers, 2019), 59–96.

75. Danielle Labbé et al., *Facing the Urban Transition in Hanoi: Recent Urban Planning Issues and Initiatives* (Montreal: INRS Centre-Urbanisation Culture Société, 2010).
76. Danielle Labbé and Clement Musil, "Periurban Land Redevelopment in Vietnam Under Market Socialism," *Urban Studies* 51, no. 6 (2013): 1146–1161.
77. Brad Evans and Julian Reid, *Resilient Life: The Art of Living Dangerously* (Malden, MA: Polity, 2014).
78. Douglas Starr, "Just 90 Companies Are to Blame for Most Climate Change," *Science*, August 25, 2016, https://www.sciencemag.org/news/2016/08/just-90-companies-are-blame-most-climate-change-carbon-accountant-says.
79. Oxfam, *Extreme Carbon Inequality: Why the Paris Climate Deal Must Put the Poorest, Lowest Emitting and Most Vulnerable People First* (Oxfam Media Briefing, 2015), https://oxfamilibrary.openrepository.com/handle/10546/582545?show=full.
80. Kenneth A Gould and Tammy L Lewis, "The Environmental Injustice of Green Gentrification: The Case of Brooklyn's Prospect Park," in *The World in Brooklyn: Gentrification, Immigration, and Ethnic Politics in a Global City*, ed. Judith DeSena and Noel Anderson (Lanham, MD: Lexington, 2012), 113–146; Andrea Chegut et al., "Supply, Demand and the Value of Green Buildings," *Urban Studies* 51, no. 1 (2014): 22–43; Cecilia Keating, "Did a Green Development Project Drive Up the Rent in a Montreal Neighbourhood," *National Observer* (Canada), January 23, 2019, https://www.nationalobserver.com/2019/01/23/features/did-green-development-project-drive-rent-montreal-neighbourhood; Isabelle Anguelovski et al., "Assessing Green Gentrification in Historically Disenfranchised Neighborhoods: A Longitudinal and Spatial Analysis of Barcelona," *Urban Geography* 39, no. 3 (2018): 458–491; Ann Dale and Lenore L. Newman, "Sustainable Development for Some: Green Urban Development and Affordability," *Local Environment* 14, no. 7 (2009): 669–681.
81. Benjamin Herazo et al., "Sustainable Development in the Building Sector: A Canadian Case Study on the Alignment of Strategic and Tactical Management," *Project Management Journal* 43, no. 2 (2012): 84–100; Benjamin Herazo and Gonzalo Lizarralde, "Understanding Stakeholders' Approaches to Sustainability in Building Projects," *Sustainable Cities and Society* 26 (2016): 240–254; Benjamin Herazo and Gonzalo Lizarralde, "The Influence of Green Building Certifications in Collaboration and Innovation Processes," *Construction Management and Economics* 33, no. 4 (2015): 279–298.
82. I have modified the names of people and organizations here to protect confidentiality.
83. Ricardo Leoto and Gonzalo Lizarralde, "Challenges in Evaluating Strategies for Reducing a Building's Environmental Impact Through Integrated Design," *Building and Environment* 155 (2019): 34–46; Ricardo Leoto and Gonzalo Lizarralde, "Challenges for Integrated Design (ID) in Sustainable Buildings," *Construction Management and Economics* 37, no. 11 (March 12, 2019): 625–642, https://www.tandfonline.com/doi/abs/10.1080/01446193.2019.1569249?journalCode=rcme20.
84. Cheryl M. Saldanha et al., "A Study of Energy Use in New York City and Leed-Certified Buildings," in *ASHRAE and IBPSA-USA SimBuild 2016—Building Performance Modeling Conference* (Salt Lake City, UT: ASHRAE and IBPSA-USA, 2016).
85. The impossibility of designing an optimal machine or building was first proposed by Herbert Simon in the book *The Sciences of the Artificial*. Simon contended that all

designers work under a condition bounded rationality—that is, a limited capacity to collect, process, and analyze data. For Simon, designers can only create a "satisficying" solution (his term). Simon was awarded the Nobel Prize in Economics in 1978.

86. Chegut et al., "Supply, Demand," 22–43.
87. David Wachsmuth et al., "Expand the Frontiers of Urban Sustainability," *Nature News*, 536, no. 7617 (2016): 391; The Atmospheric Fund (TAF), *Embodied Carbon in Construction: Policy Primer for Ontario* (Toronto: Zizzo Strategy, 2017).
88. Jennifer O'Connor et al., *Reducing Embodied Environmental Impacts of Buildings: Policy Options and Technical Infrastructure* (Ottawa: Athena Sustainable Materials Institute, 2019).
89. I have modified the architect's name here to protect their confidentiality.
90. Valérie Levée, "La Certification Passivhaus Au Québec," *Formes* 15, no. 2020 (2019): 37–38.
91. David Owen, *The Conundrum: How Scientific Innovation, Increased Efficiency, and Good Intentions Can Make Our Energy and Climate Problems Worse* (New York: Riverhead, 2012), 4.
92. One particular message in the Brundtland report refers to the idea that resources need to be managed to produce maximum yield to human beings: "most renewable resources are part of a complex and interlinked ecosystem, and maximum sustainable yield must be defined after taking into account system-wide effects of exploitation." Brundtland, *Report of the World Comission*, https://sustainabledevelopment.un.org/content/documents/5987our-common-future.pdf , 43.
93. John Livingston, *The Fallacy of Wildlife Conservation and One Cosmic Instant: A Natural History of Human Arrogance* (Toronto: McClelland & Stewart, 2013), 32.

4. Resilience: "They Say That We Must Adapt"

1. Nancy Lozano-Gracia et al., *Haitian Cities: Actions for Today with an Eye on Tomorrow* (Washington, D.C.: The World Bank—IBRD—IDA, 2017), 2, 61.
2. PMO India (@PMOIndia), "The people of Gujarat are blessed with a strong spirit of resilience. These floods will not impact the development journey of Gujarat: PM," Twitter, July 25, 2017, 10:42 a.m., https://twitter.com/PMOIndia/status/889858445368631297.
3. Michael Safi, "India Floods: 213 Killed in Gujarat as Receding Waters Reveal More Victims," *Guardian*, July 31, 2017, https://www.theguardian.com/world/2017/jul/31/india-monsoon-floods-gujarat-death-toll-over-200.
4. Michael J Watts, "Now and Then: The Origins of Political Ecology and the Rebirth of Adaptation as a Form of Thought," in *The Routledge Handbook of Political Ecology*, ed. Tom Perreault et al. (New York: Routledge, 2015), 19–42, 36.
5. Parul Sehgal, "The Profound Emptiness of 'Resilence,' " *New York Times Magazine*, December 1, 2015, https://www.nytimes.com/2015/12/06/magazine/the-profound-emptiness-of-resilience.html.

6. David Alexander, "Resilience and Disaster Risk Reduction: An Etymological Journey," *Natural Hazards and Earth System Sciences Discussions* 1, no. 13 (2013): 2707–2716.
7. United Nations Development Program, *Towards a Disaster Resilient Community in Gujarat* (New Delhi: UNDP, 2007).
8. See, for instance, Virendra Pandit, "A Resilient, Born-Again City Is Poised to Soar," *Hindu Business Line*, February 15, 2016, https://www.thehindubusinessline.com/specials/a-resilient-bornagain-city-is-poised-to-soar/article8241217.ece.
9. See, for instance, "Urban Risk Reduction and Resilience: Capacity Development for Making Cities Resilient to Disasters Training Workshop, Ahmedabad, India," UN Office for Disaster Risk Reduction, https://www.unisdr.org/campaign/resilientcities/news-events/article/57883/urban-risk-reduction-and-resilience-capacity-development-for-making-cities-resilient-to-disasters-training-workshop-ahmedabad-india.
10. Gujarat State Disaster Management Authority, *Earthquake Management Plan 2015–16* (Gandhinagar: GSDM, 2016).
11. Jonathan Pugh, "Resilience, Complexity and Post-Liberalism," *Area* 46, no. 3 (2014): 313–319.
12. I. Kelman et al., "Learning from the History of Disaster Vulnerability and Resilience Research and Practice for Climate Change," *Natural Hazards* 82, no. 1 (2016): 129–143, 137.
13. Alexander, "Resilience and Disaster Risk Reduction," 2707–2716.
14. Juergen Weichselgartner and Ilan Kelman, "Geographies of Resilience: Challenges and Opportunities of a Descriptive Concept," *Progress in Human Geography* 39, no. 3 (2015): 249–267.
15. Steve Carpenter et al., "From Metaphor to Measurement: Resilience of What to What?," *Ecosystems* 4, no. 8 (2001): 765–781.
16. A. K. Magnan et al., "Addressing the Risk of Maladaptation to Climate Change," *Wiley Interdisciplinary Reviews: Climate Change* 7, no. 5 (2016): 646–665.
17. James Lewis, "Some Realities of Resilience: A Case-Study of Wittenberge," *Disaster Prevention and Management* 22, no. 1 (2013): 48–62, 48.
18. I. Sudmeier-Rieux Karen, "Resilience—an Emerging Paradigm of Danger or of Hope?," *Disaster Prevention and Management* 23, no. 1 (2014): 67–80, 75.
19. See Marilla Steuter-Martin and Loreen Pindera, "Looking Back on the 1998 Ice Storm 20 Years Later," CBC News, January 4, 2018, https://www.cbc.ca/news/canada/montreal/ice-storm-1998-1.4469977.
20. See J. R. Minkel, "The 2003 Northeast Blackout—Five Years Later," *Scientific American*, August 13, 2008, https://www.scientificamerican.com/article/2003-blackout-five-years-later/.
21. James Barron, "The Blackout for 2003: The Overview; Power Surger Blacks Out Northeast, Hitting Cities in 8 States and Canada; Midday Shutdowns Disrupt Millions," *New York Times*, August 15, 2003, https://www.nytimes.com/2003/08/15/nyregion/blackout-2003-overview-power-surge-blacks-northeast-hitting-cities-8-states.html.

22. The character and the story are real and fully documented, but this is a modified name.
23. Alex Wilson, "The Most Resilient House in North America," Resilient Design Institute, January 19, 2015, https://www.resilientdesign.org/the-most-resilient-house-in-north-america/; Emmanuelle Waiter, "Super-résilente, hyper-éco-énergétique: La maison kénogami bat tous les records canadiens," Éco Habitation, August 30, 2013, https://www.ecohabitation.com/guides/1441/super-resiliente-hyper-eco-energetique-la-maison-kenogami-bat-tous-les-records-canadiens/.
24. James McWilliams, *Just Food: Where Locavores Get It Wrong and How We Can Truly Eat Responsibly* (New York: Hachette Book Group, 2010), 18.
25. David Owen, *The Conundrum: How Scientific Innovation, Increased Efficiency, and Good Intentions Can Make Our Energy and Climate Problems Worse* (New York: Riverhead, 2012), 73.
26. See details here: Wilson, "The Most Resilient House in North America."
27. Iain White and Paul O'Hare, "From Rhetoric to Reality: Which Resilience, Why Resilience, and Whose Resilience in Spatial Planning?," *Environment and Planning C* 32 (2014): doi:10.1068/c12117.
28. Gonzalo Lizarralde et al., "We Said, They Said: The Politics of Conceptual Frameworks in Disasters and Climate Change in Colombia and Latin America," *International Journal of Disaster Prevention and Management* 29, no. 6 (July 2020): doi: 10.1108/DPM-01-2020-0011.
29. Gonzalo Lizarralde et al., "A Systems Approach to Resilience in the Built Environment: The Case of Cuba," *Disasters* 39, no. s1 (2015): s76–s95.
30. Información comercial, "Cali fue incluida entre las 100 ciudades más resilientes del mundo," *El Pais* (Colombia), November 16, 2016, https://www.elpais.com.co/cali/fue-incluida-entre-las-100-ciudades-mas-resilientes-del-mundo.html.
31. Armando Muñiz González, *Ponencia contribución a la elevación de la resiliencia urbana de las principales ciudades de Cuba* (La Habana, Cuba: Instituto de Planificación Física, 2018).
32. UN Habitat et al., *Manual técnico de resiliencia urbana: Instrucción metodológica resiliencia urbana* (La Habana, Cuba: UNDP, 2015).
33. Eduardo Palomares Calderón, "A debate resiliencia urbana en Cuba," *Granma*, October 28, 2018, http://www.granma.cu/cuba/2018-10-28/a-debate-resiliencia-urbana-en-cuba-28-10-2018-19-10-26.
34. UNDP, *Proyecto PNUD/UE: "Incremento de la resiliencia energética de las comunidades ante eventos meteorológicos extremos a partir del uso de fuentes renovables de energía (FRE)"* (La Habana, Cuba: UNDP, 2019), 1.
35. "Nuevo proyecto para aumentar la resiliencia costera en Cuba," UNDP, February 25, 2020, https://www.cu.undp.org/content/cuba/es/home/presscenter/articles/2019/resilienciacosteraenCuba.html.
36. Morgan Scoville-Simonds et al., "The Hazards of Mainstreaming: Climate Change Adaptation Politics in Three Dimensions," *World Development* 125 (2020): 1–10, 1.
37. Weichselgartner and Kelman, "Geographies of Resilience," 262.

38. Sijla Klepp and Libertad Chavez-Rodriguez, eds., *A Critical Approach to Climate Change Adaptation: Routledge Advances in Climate Change Research* (New York: Routledge, 2018), 3.
39. Khaled Al-Kassimi, "The Logic of Resilience as Neoliberal Governmentality Informing Hurricane Katrina and Hurricane Harvey—Is Cuban Resilience Strategy (Crs) an Alternative?," *Ottawa Law Review* (2019): DOI: 10.18192/potentia.v10i0.4509.
40. Greg Bankoff, "Remaking the World in Our Own Image: Vulnerability, Resilience and Adaptation as Historical Discourses," *Disasters*, 43, no. 2 (2019): 221–239.
41. T. J. Marsh et al., *The Winter Floods of 2015/2016 in the UK—A Review* (Wallingford, UK.: Centre for Ecology & Hydrology, 2016), https://www.ceh.ac.uk/sites/default/files/2015-2016%20Winter%20Floods%20report%20Low%20Res.pdf.
42. Samuel Osborne, "10,000 UK Homes Built on Flood Plains Each Year," *Independent* (UK), December 28, 2015, https://www.independent.co.uk/news/uk/home-news/10000-uk-homes-built-on-flood-plains-each-year-a6788816.html.
43. David Crichton, "Toward an Integrated Approach to Managing Flood Damage," *Building Research & Information* 33, no. 3 (2005): 293–299.
44. Fiona Harvey, "Nine Tenths of England's Floodplains Not Fit for Purpose, Study Finds," *Guardian*, May 31, 2017, https://www.theguardian.com/environment/2017/jun/01/englands-90-floodplains-not-fit-for-purpose-study-finds.
45. Committee on Climate Change—Adaptation Sub-Committee, *Climate Change—Is the Uk Preparing for Flooding and Water Scarcity?* (London: Committee on Climate Change, 2012).
46. Committee on Climate Change, *Climate Change—Is the Uk Preparing for Flooding and Water Scarcity?*
47. Simon Mclean and Paul Watson, "A Practical Approach to Development of Housing on Floodplain Land in the UK," *Journal of Building Appraisal* 4, no. 4 (2009): 311–320.
48. Gonzalo Lizarralde et al., "The Diversity of Governance Approaches in the Face of Resilience," in *Governance of Risk, Hazards and Disasters: Trends in Theory and Practice*, ed. Giuseppe Forino et al. (New York, London: Routledge, 2018); Gonzalo Lizarralde et al., *Tensions and Complexities in Creating a Sustainable and Resilient Built Environment: Achieving a Turquoise Agenda in the UK* in *Proceedings of the 5th International Disaster and Risk Conference* (Davos, Switzerland: IDRC, 2014); Gonzalo Lizarralde et al., "Sustainability and Resilience in the Built Environment: The Challenges of Establishing a Turquoise Agenda in the UK," *Sustainable Cities and Society* 15 (2015): 96–104; J. Fisher et al., "Urban Resilience and Sustainability: The Role of a Local Resilience Forum in England," in *Disaster Management: Enabling Resilience*, ed. A. J. Masys (New York: Springer, 2014), 91–107; Ksenia Chmutina et al., "Unpacking Resilience Policy Discourse," *Cities* 58 (2016): 70–79;
49. Lizarralde et al., "Sustainability and Resilience in the Built Environment." Subsequent quotes in this section, unless otherwise noted, are to this article.
50. Ksenia Chmutina et al., "The Reification of Resilience and the Implications for Theory and Practice," in *Realcorp 2014*, ed. M. Schrenk et al. (Vienna, Austria: RealCorp 2014, 2014).

51. Watts, "Now and Then," 40.
52. Brad Evans and Julian Reid, *Resilient Life: The Art of Living Dangerously* (Malden, MA: Polity, 2014).
53. Bankoff, "Remaking the World in Our Own Image."
54. Watts, "Now and Then," 41.
55. Ilan Kelman et al., eds., *The Routledge Handbook of Disaster Risk Reduction Including Climate Change Adaptation.* (London: Routledge, 2017), 52.
56. Isabelle Anguelovski et al., "Equity Impacts of Urban Land Use Planning for Climate Adaptation: Critical Perspectives from the Global North and South," *Journal of Planning Education and Research* 36, no. 3 (2016): 333–348.
57. Lizzie Yarina, "Your Sea Wall Won't Save You: Negotiating Rhetorics and Imaginaries of Climate Resilience," *Places*, March 2018, https://placesjournal.org/article/your-sea-wall-wont-save-you/?cn-reloaded=1&cn-reloaded=1.
58. A. Leiserowitz et al., *Climate Change in the American Mind: April 2019*, ed. Yale University and George Mason University (New Haven, CT: Yale Program on Climate Change Communication, 2019), doi:10.17605/OSF.IO/CJ2NS.
59. Klepp and Chavez-Rodriguez, *A Critical Approach*, 14.
60. Katrina Brown, *Resilience, Development and Global Change* (New York: Routledge, 2016), 54.

5. Participation: "They Want Us to Participate in the Construction of I-Don't-Know-What"

1. UNDRO, *Shelter After Disaster: Guidelines for Assistance* (New York: United Nations, 1982), 55.
2. Gro Harlem Brundtland, *Report of the World Commission on Environment and Development: "Our Common Future"* (New York: United Nations, 1987), 45.
3. UNISDR, *The Hyogo Framework for Action 2005–2015* (Geneva: United Nations International Strategy for Disaster Reduction, UNISDR, 2005), 13.
4. Anamika Barua et al., *Climate Change Governance and Adaptation: Case Studies from South Asia* (London: CRC Press, 2018).
5. Joydeep Gupta, "Book Review: Climate Adaptation Impossible Without Communtiy Participation," review of *Climate Change Governance and Adaptation: Case Studies from South Asia*, ed. Anamika Barua, Vishal Narain, and Sumit Vij, TheThirdPole.Net, May 21, 2019, https://www.thethirdpole.net/en/2019/05/21/book-review-climate-adaptation-impossible-without-community-participation/.
6. Find the debate here: "4th Debate: Is Public Participation Really the Key to Success for Urban Projects and Initiatives Aimed at Disaster Risk Reduction?," Vulnerability, Resiliance, and Post-Disaster Reconstruction International Debates, https://odde-bates.com/4-fourth-debate/.
7. Gonzalo Lizarralde et al., "Framing Responses to Post-Earthquake Haiti: How Representations of Disasters, Reconstruction and Human Settlements Shapes Resilience,"

International Journal of Disaster Resilience and the Built Environment 4, no. 1 (2011): 43–57.

8. G. Lizarralde et al., "L'habitat (Chapitre III)," in *Perspectives De Développement De L'aire Métropolitaine De Port-Au-Prince, Horizon 2030*, ed. J. Goulet et al. (Montreal: UQAM-Université de Montréal, 2018), 149–192.
9. Thomas Perreault and Patricia Martin, *Geographies of Neoliberalism in Latin America* (London: Sage Publications, 2005); Jonathan Joseph, "Resilience as Embedded Neoliberalism: A Governmentality Approach," *Resilience* 1, no. 1 (2013): 38–52; Carlos Arze and Tom Kruse, "The Consequences of Neoliberal Reform," *NACLA Report on the Americas* 38, no. 3 (2004): 23–28.
10. Robert Fatton Jr., *Haiti: Trapped in the Outer Periphery* (Boulder, CO: Lynne Rienner, 2014).
11. Jean-Germain Gros, "Haiti: The Political Economy and Sociology of Decay and Renewal," *Latin American Research Review* 35, no. 3 (2000): 211–226.
12. Alex Dupuy, "Disaster Capitalism to the Rescue: The International Community and Haiti After the Earthquake," *NACLA Report on the Americas* 43, no. 4 (2010): 14–19.
13. See Michael R. Gordon, "Standoff in Haiti: U.S. Strategy; U.S. Troops Stage Military Exercise with Eye on Haiti," *New York Times*, July 7, 1994, https://www.nytimes.com/1994/07/07/world/standoff-haiti-us-strategy-us-troops-stage-military-exercise-with-eye-haiti.html.
14. See, for instance, Kevin Sullivan and Rosalind S. Heiderman, "How the Clintons' Haiti Development Plans Succeed—and Disappoint," *Washington Post*, March 20, 2016, https://www.washingtonpost.com/politics/how-the-clintons-haiti-development-plans-succeed—and-disappoint/2015/03/20/0ebae25e-cbe9-11e4-a2a7-9517a3a70506_story.html; see also Janet Reitman, "Beyond Relief: How the World Failed Haiti," *Rolling Stone*, August 4, 2011, https://www.rollingstone.com/politics/politics-news/beyond-relief-how-the-world-failed-haiti-242928/.
15. Find a pertinent article here: Reitman, "Beyond Relief." See video here: " 'We Made a Devil's Bargain': Fmr. President Clinton Apologizes for Trade Politics That Destroyed Haitian Rice Farming," Democracy Now!, video, 38:17, April 1, 2010, https://www.democracynow.org/2010/4/1/clinton_rice.
16. Mark Schuller, "Gluing Globalization: NGOs as Intermediaries in Haiti," *PoLAR: Political and Legal Anthropology Review* 32, no. 1 (2009): 84–104.
17. Dupuy, "Disaster Capitalism to the Rescue," 211–226.
18. Find more information on this subject here: "Projet," PRCU: Programme de Recherche dans le Champ de l'Urbain, http://www.repertoiregrif.umontreal.ca/prcu/.
19. Laura Puertas and Jon Elsen, "Earthquake in Peru Kills Hundreds," *New York Times*, August 16, 2007, https://www.nytimes.com/2007/08/16/world/americas/16cnd-peru.html.
20. Samir Elhawary and Gerardo Castillo, *The Role of the Affected State: A Case Study on the Peruvian Earthquake Response* (Lima, Peru: CIES- Economic and Social Research Consortium, 2008), 17.

21. Jess Rapp, "2007 Peru Earthquake 10 Years On: Is Peru Prepared for the Next Big One?," Peru Reports, November 24, 2017, https://perureports.com/2007-peru-earthquake-10-years-peru-prepared-next-one/6197/.
22. See the official plan for the 2017 reconstruction here: https://busquedas.elperuano.pe/normaslegales/decreto-supremo-que-aprueba-el-plan-de-la-reconstruccion-al-decreto-supremo-n-091-2017-pcm-1564235-1/
23. I have written the following sentences from different pieces of empirical research. They are not actual quotes. The message they express illustrate the findings of my empirical case studies.
24. Patricia A Wilson, "Deliberative Planning for Disaster Recovery: Re-Membering New Orleans," *Journal of Public Deliberation* 5, no. 1 (2009): 1; Isabelle Thomas and James Amdal, "Stakeholder Participation in Post-Disaster Reconstruction Programmes—New Orleans' Lakeview: A Case Study," in *Rebuilding After Disasters: From Emergency to Sustainability*, ed. Gonzalo Lizarralde et al. (New York: Routledge, 2009), 120–142; Lewis D Hopkins, Divya Chandrasekhar, Kanako Iuchi, and Robert Olshansky, *How Do Organizations Use Plans in Urban Development? A Case from New Orleans*. 2016. doi:10.13140/RG.2.2.26894.36161.
25. One of the first scholars to explain this was Sherry Arnstein, an American intellectual and consultant. See Sherry R. Arnstein, "A Ladder of Citizen Participation," *Journal of the American Planning Association* 35, no. 4 (1969): 216–224.
26. John F. C. Turner and Robert Fichter, *Freedom to Build: Dweller Control of the Housing Process* (New York: Macmillan, 1972), 174.
27. Richard Harris, "A Double Irony: The Originality and Influence of John F. C. Turner," *Habitat International* 27, no. 2 (2003): 245–269.
28. Find a detailed explanation of housing policies in Gonzalo Lizarralde, *The Invisible Houses: Rethinking and Designing Low-Cost Housing in Developing Countries* (London: Routledge, 2014).
29. Rod Burgess, "Petty Commodity Housing or Dweller Control? A Critique of John Turner's Views on Housing Policy," *World Development* 6, no. 9 (1978): 1105–1133.
30. Lizarralde, *The Invisible Houses*.
31. Ramin Keivani and Edmundo Werna, "Refocusing the Housing Debate in Developing Countries from a Pluralist Perspective," *Habitat International* 25, no. 2 (2001): 191–208.
32. Jennifer Rietbergen-McCracken, ed., *Participation in Practice: The Experience of the World Bank and Other Stakeholders* (Washington, D.C.: World Bank, 1996).
33. S. R. Arnstein, "A Ladder of Citizen Participation," *Journal of the American Planning Association* 35, no. 4 (1969): 216–224.
34. L. Blondiaux and Y. Sintomer, "L'impératif délibératif," *Politix* 15, no. 57 (2002): 17–35.
35. Loïc Blondiaux, *Le nouvel esprit de la démocratie* (Paris: Seuil, Collection La République des idées, 2008); Loïc Blondiaux, "La démocratie participative, sous conditions et malgré tout," *Mouvements*, no. 2 (2007): 118–129.
36. Katherine V. Gough and Peter Kellett, "Housing Consolidation and Home-Based Income Generation: Evidence from Self-Help Settlements in Two Colombian

Cities," *Cities* 18, no. 4 (2001): 235–247; Alan Gilbert, "Financing Self-Help Housing: Evidence from Bogotá, Colombia," *International Planning Studies* 5, no. 2 (2000): 165–190.

37. B. Ferguson and J. Navarrete, "New Approaches to Progressive Housing in Latin America: A Key to Habitat Programs and Policy," *Habitat International* 27, no. 2 (2003): 309–323.
38. Gonzalo Lizarralde and David Root, "The Informal Construction Sector and the Inefficiency of Low Cost Housing Markets," *Construction Management and Economics* 26, no. 2 (2008): 103–113.
39. Dhouha Bouraoui and Gonzalo Lizarralde, "Centralized Decision Making, Users' Participation and Satisfaction in Post-Disaster Reconstruction: The Case of Tunisia," *International Journal of Disaster Resilience in the Built Environment* 4, no. 2 (2013): 145–167.
40. Samuel Paul, *Community Participation in Development Projects: The World Bank Experience* (Washington, D.C.: World Bank, 1987) 12, 10.
41. Rietbergen-McCracken, *Participation in Practice*, 2.
42. United Nations Habitat, *The Instanbul Declaration and the Habitat Agenda* (Nairobi: United Nations Human Settlements Program, 1996), 38.
43. UNCHS—Habitat, *Cities in a Globalizing World: Global Report on Human Settlements 2001* (London: Earthscan, 2001), 84.
44. Marisa B. Guaraldo Choguill, "A Ladder of Community Participation for Underdeveloped Countries," *Habitat International* 20, no. 3 (1996): 431–444; Charles L. Choguill, "The Search for Policies to Support Sustainable Housing," *Habitat International* 31, no. 1 (2007): 143–149.
45. UNDP, *Guidelines for Community Participation in Disaster Recovery* (n.p.: United Nations Development Programme, 2020), 8.
46. Federation of Canadian Municipalities, *Local Government and Poverty Reduction: Report for Instanbul+5* (Ottawa: Federation of Canadian Municipalities, 2001), 19.
47. James Lewis and Ilan Kelman, "Places, People and Perpetuity: Community Capacities in Ecologies of Catastrophe," *ACME* 9, no. 2 (2010), 208.
48. C. H. Davidson et al., "Truths and Myths About Community Participation in Post-Disaster Housing Projects," *Habitat International* 31, no. 1 (2007): 100–115.
49. Michael J. Sandel, *Justice: What's the Right Thing to Do?* (London: Penguin, 2010).
50. Roberto Merrill, "Comment Un État Libéral Peut-Il Être À La Fois Neutre Et Paternaliste?," *Raisons politiques*, no. 4 (2011): 15–40.
51. Michael J. Sandel, *What Money Can't Buy: The Moral Limits of Markets* (New York: Farrar, Strauss and Giroux, 2012).
52. Farnçois Cardinal, "Pas dans mon parc," *La Presse*, September 22, 2011, http://www.cyberpresse.ca/debats/editorialistes/francois-cardinal/201109/22/01-4450199-pas-dans-mon-parc.php.
53. Eric R. A. N. Smith and Marisela Marquez, "The Other Side of the NIMBY Syndrome," *Society & Natural Resources* 13, no. 3 (2000): 273–280; Timothy A. Gibson, "Nimby and the Civic Good," *City & Community* 4, no. 4 (2005): 381–401; Michael

Dear, "Understanding and Overcoming the NIMBY Syndrome," *Journal of the American Planning Association* 58, no. 3 (1992): 288–300; See, for instance: Pascale Breton, "Pas d'école dans ma cour," *La Presse*, July 5, 2014, https://www.lapresse.ca/debats/201407/04/01-4781228-pas-decole-dans-ma-cour.php.

54. Rosanvallon Pierre, *La Contre-Démocratie: La Politique À L'âge De La Défiance* (Paris: Le Seuil, 2006); Marcel Gauchet, "Crise dans la Démocratie," *La revue lacanienne*, no. 2 (2008): 59–72.
55. Richard A Posner, *The Crisis of Capitalist Democracy* (Cambridge, MA: Harvard University Press, 2010); On populist politicians, see, for instance, Paul Lewis et al., "Revealed: The Rise and Rise of Populist Rhetoric," *Guardian*, March 6, 2019, https://www.theguardian.com/world/ng-interactive/2019/mar/06/revealed-the-rise-and-rise-of-populist-rhetoric.
56. Local leaders sometimes specify that they do not oppose density but the construction of highrise buildings. I refer here to the density of neighborhoods and territories rather than the occupation of single blocks or plots.
57. See, for instance, Yasminah Beebeejaun, "The Participation Trap: The Limitations of Participation for Ethnic and Racial Groups," *International Planning Studies* 11, no. 1 (2006): 3–18.
58. UNDP, *Guidelines for Community Participation*.

6. Innovation: "We Need Something Really Innovative, They Said"

1. "Haiti: After the Quake," *Montreal Gazette*, 2011, http://www.montrealgazette.com/news/haiti-quake/index.html.
2. See Kevin Sullivan and Rosalid S. Helderman, "How the Clintons' Haiti Development Plans Succeed—and Disappoint," *Washington Post*, March 20, 2015, https://www.washingtonpost.com/politics/how-the-clintons-haiti-development-plans-succeed—and-disappoint/2015/03/20/0ebae25e-cbe9-11e4-a2a7-9517a3a70506_story.html.
3. See Sullivan and Helderman, "How the Clintons' Haiti Development."
4. Find a pertinent article at Rene Bruemmer, "Homes for Haiti: The Plastic House," *Montreal Gazette*, July 8, 2011, http://www.montrealgazette.com/homes+haiti+plastic+house/5073804/story.html.
5. The full report can be found at Christian Werthmann, Phil Thompson, Dan Weissman, and Anya Brickman Raredon, eds., *Designing Process: Exemplar Community Pilot Project—Zoranje, Port au Prince, Haiti*, 2nd ed. (Cambridge, MA: Harvard University Graduate School of Design; MIT School of Architecture and Planning, 2011), 140, 155, http://web.mit.edu/colab/pdf/papers/Designing_Process_Haiti.pdf.
6. Gonzalo Lizarralde, *The Invisible Houses: Rethinking and Designing Low-Cost Housing in Developing Countries* (London: Routledge, 2014).
7. Vijaya Ramachandran and Julie Walz, "Haiti: Where Has All the Money Gone?," *Journal of Haitian Studies* 21, no. 1 (2015): 26–65, 27.

8. See Joaquín Villanueva, "Intervention—"Beyond Disaster Capitalism: Dismantling the Infrastructure of Extraction in Puerto Rico's Neo-Plantation Economy," *Antipode Online*, June 25, 2019, https://antipodeonline.org/2019/06/25/beyond-disaster-capitalism/?fbclid=IwAR3NnnJaw100O37eal7ucBjXXdm91k-sbCVsGNflOgd8uKdILAGWQY-sEokw.
9. R. Forrest and P. Williams, "Commodification and Housing: Emerging Issues and Contradictions," *Environment and Planning A: Economy and Space* 16, no. 9 (1984): 1163–1180.
10. Jennifer Duyne Barenstein et al., *Safer Homes, Stronger Communities: A Handbook for Reconstruction after Natural Disasters* (Washington, D.C.: World Bank, 2010).
11. Jennifer E. Duyne Barenstein and Esther Leemann, *Post-Disaster Reconstruction and Change: Communities' Perspectives* (London: CRC Press, 2012).
12. Rohit Jigyasu, "Appropriate Technology for Post-Disaster Reconstruction," in *Rebuilding after Disasters: From Emergency to Sustainability*, ed. Gonzalo Lizarralde et al. (New York: Taylor and Francis, 2010), 49–69.
13. Gro Harlem Brundtland, *Report of the World Commission on Environment and Development: "Our Common Future"* (New York: United Nations, 1987), n.p.
14. Steven Pinker, *Enlightenment Now: The Case for Reason, Science, Humanism, and Progress* (New York: Viking, 2018), 329.
15. David E. Alexander, "The L'Aquila Earthquake of 6 April 2009 and Italian Government Policy on Disaster Response," *Journal of Natural Resources Policy Research* 2, no. 4 (2010): 325–342.
16. Giuseppe Forino, "Disaster Recovery: Narrating the Resilience Process in the Reconstruction of L'Aquila (Italy)," *Geografisk Tidsskrift-Danish Journal of Geography* 115, no. 1 (2015): 1–13.
17. See, for instance, "C.A.S.E. Project: L'Aquila (Italy)," Sistem Costruzioni, https://www.sistem.it/en/realizations/residential-buildings/condominiums-multi-storey-buildings-and-social-housing/c-a-s-e-project-laquila-italy/.
18. Jan-Jonathan Bock, "The Second Earthquake: How the Italian State Generated Hope and Uncertainty in Post-Disaster L'Aquila," *Journal of the Royal Anthropological Institute* 23, no. 1 (2017): 61–80.
19. Bock, "The Second Earthquake."
20. Giuseppe Forino and Fabio Carnelli, "Introduction to the Special Issue 'the L'Aquila Earthquake 10 Years on (2009–2019): Impacts and State-of-the-Art,' " *Disaster Prevention and Management: An International Journal* 28, no. 4 (2019): 414–418, 414.
21. David Alexander, "An Evaluation of Medium-Term Recovery Processes after the 6 April 2009 Earthquake in L'Aquila, Central Italy," *Environmental Hazards* 12, no. 1 (2013): 60–73, 60.
22. Francesca Fois and Giuseppe Forino, "The Self-Built Ecovillage in L'Aquila, Italy: Community Resilience as a Grassroots Response to Environmental Shock," *Disasters* 38, no. 4 (2014): 719–739; Diana Contreras et al., "Spatial Connectivity as a Recovery Process Indicator: The L'Aquila Earthquake," *Technological Forecasting and Social Change* 80, no. 9 (2013): 1782–1803.

23. Diana Contreras et al., “Lack of Spatial Resilience in a Recovery Process: Case L’Aquila, Italy,” *Technological Forecasting and Social Change* 121 (August 2017): 76–88, 85.
24. Diana Contreras et al., “Measuring the Progress of a Recovery Process After an Earthquake: The Case of L’Aquila, Italy,” *International Journal of Disaster Risk Reduction* 28, no. (2018): 450–464.
25. Find details of the INCIPICT project here: http://incipict.univaq.it/.
26. OECD, *Policy Making after Disasters: Helping Regions Become Resilient—the Case of Post-Earthquake Abruzzo* (Paris: OECD Publishing, 2013).
27. See Rete8, “L’Aquila - Smart City, Forum de ‘Il Centro,’ ” YouTube video, 2:15, November 30, 2017, https://www.youtube.com/watch?v=BDIytQtz_io.
28. C. Antonelli et al., *The City of L’Aquila as a Living Lab: The Incipict Project and the 5G Trial*, in *2018 IEEE 5G World Forum (5GWF)*, 2018, 177.
29. Enzo Falco et al., “Smart City L’Aquila: An Application of the ‘Infostructure’ Approach to Public Urban Mobility in a Post-Disaster Context,” *Journal of Urban Technology* 25, no. 1 (2018): 99–121.
30. Vincenzo Gattulli et al., *Distributed Structural Monitoring for a Smart City in a Seismic Area*, in *Key Engineering Materials* (Zurich: Trans Tech, 2015).
31. See “Smart Clean Air City L’Aquila,” Is Clean Air, https://www.iscleanair.com/wp/2018/10/05/smart-clean-air-city-laquila-2/?lang=en.
32. Gattulli et al., Distributed Structural Monitoring,” 133
33. See “Premio ‘Smart Communities,’ tra i vincitori il Comune dell’Aquila,” Casa & Clima, October 28, 2015, https://www.casaeclima.com/ar_24732_14811_premio-smart-communities-vincitori-comune-aquila-smart-city.html. See also “L’Aquila, Varese, Treviso e Napoli I vincitori del Premio Smart City,” October 28, 2013, http://www.rinnovabili.it/smart-city/premio-smart-city-smau-vincitori-2013/.
34. Barbara Lucini, “Lifestyle Practices and Cultural Survival after L’Aquila Earthquake (Italy, 2009),” in *Disaster’s Impact on Livelihood and Cultural Survival: Losses, Opportunities and Mitigation*, ed. Michele Companion, 269–280 (New York: CRC Press, 2015).
35. Giorgos Koukoufikis, “Post-Disaster Redevelopment and the ‘Knowledge City’: Limitations of an Urban Imaginary in L’Aquila,” *Disaster Prevention and Management: An International Journal* 28, no. 4 (2019): 474–486, 482, 483.
36. Suzanne Wilkinson et al., “Reconstruction Following Earthquake Disasters,” *Encyclopedia of Earthquake Engineering*, ed. Michael Beer et al. (New York: Springer, 2015), 1–11.
37. See Kurt Bayer, “Christchurch Rebuild Plan Revealed,” *New Zealand Herald*, July 30, 2012, https://www.nzherald.co.nz/nz/news/article.cfm?c_id=1&objectid=10823289.
38. Lukas Marek et al., “Shaking for Innovation: The (Re)Building of a (Smart) City in a Post Disaster Environment,” *Cities* 63 (2017): 41–50.
39. The Rockefeller Foundation and Christchurch city council resilience project team, *Resilient Greater Christchurch* (The Rockefeller Foundation, 2016), http://greaterchristchurch.org.nz/assets/Documents/greaterchristchurch/Resilient/Resilient-Greater-Christchurch-Plan.pdf.

40. Resilient Greater Christchurch.
41. See "Christchurch's Sensing City Project Shelved," *Idealog*, September 4, 2015, https://idealog.co.nz/tech/2015/09/christchurchs-sensing-city-project-shelved. See also Charlotte Fernández, "World-First Sensing City Project Launch," Arup, September 2, 2013, https://www.arup.com/news-and-events/worldfirst-sensing-city-project-launched.
42. IBM Corporate Citizenship & Corporate Affairs, *IBM's Smarter Cities Challenge Report Christchurch* (New York: IBM, 2013), 1.
43. L. Marek et al., "Real-Time Environmental Sensors to Improve Health in the Sensing City," *International Archives of Photogrammetry, Remote Sensing and Spatial Information Sciences* 41 (2016): 729.
44. Found at "About," Smart Christchurch, https://smartchristchurch.org.nz/about/.
45. IBM Corporate Citizenship & Corporate Affairs, *IBM's Smarter Cities*, n.p.
46. See "This Smart City in New Zealand Is Developing Around Resiliency Using the Power of Data," SmartCity.Press, October 16, 2018, https://www.smartcity.press/christchurch-smart-initiatives/.
47. Michael Wright, "Gerry Brownlee Exits Christchurch a Controversial, Contrary Figure," Stuff, April 24, 2017, https://www.stuff.co.nz/national/politics/91883859/gerry-brownlee-exits-christchurch-a-controversial-contrary-figure.
48. See Belinda McCammon, "CERA Failed to Maintain 'Momentum'—Report," RNZ (Radio New Zealand), February 8, 2017, https://www.rnz.co.nz/news/national/324077/cera-failed-to-maintain-%27momentum%27-report.
49. Controller and Auditor General, *Canterbury Earthquake Recovery Authority: Assessing Its Effectiveness and Efficiency* (Wellington, New Zealand: Controller and Auditor General New Zealand, 2017), 4, https://www.oag.govt.nz/2017/cera/docs/cera.pdf.
50. Marek et al., "Shaking for Innovation," 48.
51. See Idealog, "Christchurch's Sensing City Project Shelved." See also Richard MacManus, "Smart Cities Encounter," *Newsroom*, May 8, 2017, https://www.newsroom.co.nz/2017/05/07/24757/we-built-smart-cities-on-rock-and-roll; and Jonathan Underhill, "Pressing Need for Basic Infrastructure Overtakes Christchurch 'Smart City' Plan," NBR, September 1, 2015, https://www.nbr.co.nz/article/pressing-need-basic-infrastructure-overtakes-christchurch-smart-city-plan-b-178126.
52. Evaluation Consult, *Evaluation of the Smart Cities Programme 2015–2016—Land Information New Zealand* (Christchurch, New Zealand: Evaluation Consult, 2016), 26.
53. See Doug Peeples, "Why Christchurch Is Betting So Heavily on Sensors," Smart Cities Council Australia New Zealand, April 26, 2017, https://anz.smartcitiescouncil.com/article/why-christchurch-betting-so-heavily-sensors.
54. See Jordana Goldman, "Five Things We Learned from Naomi Klein's *The Battle for Paradise* Launch," *Now Toronto*, June 28, 2018, https://nowtoronto.com/culture/books/naomi-klein-battle-for-paradise/.
55. Canadian Center for Economic Analysis and Canadian Urban Institute, *Toronto Housing Market Analysis: From Insight To Action* (Toronto: Canadian Center for Economic Analysis, 2019).

56. Laura Bliss, "A Big Master Plan for Google's Growing Smart City," Bloomberg CityLab, June 25,2019, https://www.bloomberg.com/news/articles/2019-06-25/toronto-s-alphabet-powered-smart-city-is-growing.
57. Bliss, "A Big Master Plan."
58. As a matter of fact, this list is pretty much the same I heard in a conference I attended in the early 2000s. The speaker was Moshe Safdie, a celebrated Canadian architect. Safdie was not talking about the future of cities, though. He was talking about the ideas he developed in of series of projects he designed in the 1960s, including the famous residential project Habitat II in Montreal.
59. Jessica Patton, "Sidewalk Lab Announces It's No Longer Pursuing Toronto Waterfront Development," *Global News* (Canada), May 7, 2020, https://globalnews.ca/news/6915490/sidewalk-labs-toronto-waterfront-development/.
60. The quote was cited here: Patton, "Sidewalk Lab."
61. See Ajnand Giridharadas, "The New Elite's Phoney Crusade to Save the World—Without Changing Anything," *Guardian*, January 22, 2019, https://www.theguardian.com/news/2019/jan/22/the-new-elites-phoney-crusade-to-save-the-world-without-changing-anything.
62. Paul Hawken, ed., *Drawdown: The Most Comprehensive Plan Ever Proposed to Reverse Global Warming* (New York: Penguin, 2017). The quote is from their website, Project Drawdown, https://drawdown.org/.
63. Faye Duchin, "An Ambitious Plan to Leverage Existing Solutions to Global Warming Is Short on Analytic Rigor," *Science*, May 23, 2017, https://blogs.sciencemag.org/books/2017/05/23/drawdown/.
64. See Andrea Thompson, "Technology Won't Stop Global Warming, Economists Say," *Live Science*, November 19, 2007, https://www.livescience.com/4722-technology-won-stop-global-warming-economists.html.
65. Ozzie Zehner, *Green Illusions: The Dirty Secrets of Clean Energy and the Future of Environmentalism* (Lincoln: University of Nebraska Press, 2012), xvi.
66. Find the argument here: Charlie Sweatpants, "Quick Climate Book: Green Illusions by Ozzie Zehner," *Medium*, April 1, 2017, https://medium.com/@CSweatpants/quick-climate-book-green-illusions-by-ozzie-zehner-f24deed47a.
67. Zehner, *Green Illusions*, 340.
68. Zehner, *Green Illusions*, xvi.
69. See, for instance, Tom Zeller Jr., "Ozzie Zehner's 'Green Illusions' Ruffles Feathers," *Huffpost*, December 6, 2017, https://www.huffpost.com/entry/ozzie-zehner-green-illusions_b_1710382.
70. See Steve Horn, "Interview: Power Shift Away from Green Illusions," Truthout, April 8, 2013, https://truthout.org/articles/power-shift-away-from-green-illusions/.
71. Chandran Nair, *The Sustainable State: The Future of Government, Economy, and Society* (San Francisco: Berrett-Koehler, 2018), 504.
72. See, for instance, Zeller, "Ozzie Zehner's 'Green Illusions' Ruffles Feathers."
73. Robert Muir-Wood, *The Cure for Catastrophe: How We Can Stop Manufacturing Natural Disasters* (Philadelphia: Basic Books, 2015), 4032, 4034.

74. David Owen, *The Conundrum: How Scientific Innovation, Increased Efficiency, and Good Intentions Can Make Our Energy and Climate Problems Worse* (New York: Riverhead, 2012), 248, 249, 124.
75. Jean-Jacques Terrin, interview with author, November 2016.
76. See a report here: International Council of Shopping Centers, "Inside-Mexicos-First-Leed-Certified-Mall." https://www.icsc.com/news-and-views/icsc-exchange/inside-mexicos-first-leed-certified-mall.
77. "NREL Report Firms Up Land-Use Requirements of Solar," NREL, press release, July 20, 2013, https://www.nrel.gov/news/press/2013/2269.html; and also: "Land-Use Requirements of Solar," *Science Daily*, August 6, 2013, https://www.sciencedaily.com/releases/2013/08/130806145537.htm.
78. Benjamin K. Sovacool, *Contesting the Future of Nuclear Power: A Critical Global Assessment of Atomic Energy* (Singapore: World Scientific, 2011), 232.
79. U.S. Department of Energy—Energy Efficiency and Renewable Energy, *PV FAQs* (Washington, D.C.: U.S. Department of Energy—Energy Efficiency and Renewable Energy, 2004), 1.
80. Tom Cardoso and Matt Lundy, "Airbnb Likely Removed 31,000 Homes from Canada's Rental Market, Study Finds," *Globe and Mail* (Toronto), June 20, 2019, https://www.theglobeandmail.com/canada/article-airbnb-likely-removed-31000-homes-from-canadas-rental-market-study/.
81. Nicole Brockbank, "Airbnbs Have 'Serious Impact' on Toronto Housing Despite Making Up Less Than 1% of Dwellings, Expert Says," CBC News, May 1, 2019, https://www.cbc.ca/news/canada/toronto/airbnb-listing-data-toronto-1.5116941.
82. Kyle Barron et al., "The Effect of Home-Sharing on House Prices and Rents: Evidence from Airbnb," *Marketing Science* 40, no. 1 (January–February 2021), https://pubsonline.informs.org/doi/10.1287/mksc.2020.1227. .
83. Daniel Guttentag, "What Airbnb Really Does to a Neighbourhood," BBC News, August 2018, https://www.bbc.com/news/business-45083954.
84. Zach Dubinsky and Valérie Ouellet, "Who's Behind the Smiling Faces of Some Airbnb Hosts? Multimillion-Dollar Corporations," CBC News, April 30, 2019, https://www.cbc.ca/news/business/biggest-airbnb-hosts-canada-corporations-1.5116103.
85. Calestous Juma, *Innovation and Its Enemies: Why People Resist New Technologies* (New York: Oxford University Press, 2016), 7, 5.
86. Juma, *Innovation and Its Enemies*, 6.
87. Pinker, *Enlightenment Now*, 152.
88. In *Enlightenment Now*, Pinker manages to neither dismiss nor endorse this particular strategy of climate engineering.
89. Joseph Schumpeter, "Creative Destruction," *Capitalism, Socialism and Democracy* 825 (1942): 82–85.
90. See "World Population Projected to Reach 9.8 Billion in 2050, and 11.2 Billion in 2100," UN Department of Economic and Social Affairs, June 21, 2017, https://www.un.org/development/desa/en/news/population/world-population-prospects-2017.html.

7. Decision-Making: "We Want to Be Able to Make Our Own Decisions"

1. The Netflix series (2015–2017).
2. "Villatina: Radiografía de un desastre en Medellín," *El Espectador* (Bogota), September 27, 2017, https://www.elespectador.com/noticias/nacional/antioquia/villatina-radiografia-de-un-desastre-en-medellin-articulo-715372.
3. "Tras la tragedia en Medellín, temen más víctimas en zonas de alto riesgo por el invierno," *Semana*, May 28, 2007, https://www.semana.com/on-line/articulo/tras-tragedia-medellin-temen-mas-victimas-zonas-alto-riesgo-invierno/86247-3.
4. "El pobre siempre paga," *Semana*, June 12, 2010, https://www.semana.com/nacion/articulo/el-pobre-siempre-paga/117887-3.
5. Sibylla Brodzinsky, "From Murder Capital to Model City: Is Medellín's Miracle Show or Substance?," *Guardian*, April 17, 2014, https://www.theguardian.com/cities/2014/apr/17/medellin-murder-capital-to-model-city-miracle-un-world-urban-forum.
6. See detailed information at "Medellín Metrocables," Gondola Project: A Cable-Propelled Transit Primer, SCJ Alliance, http://gondolaproject.com/medellin/.
7. "City of the Year: Medellín," *Wall Street Journal*, March 1, 2013, http://online.wsj.com/ad/cityoftheyear.
8. Victoria Burnett, "New Mexico City, Cable Car Lets Commuters Glide Over Traffic," *New York Times*, December 28, 2016, https://www.nytimes.com/2016/12/28/world/americas/mexico-city-mexicable.html.
9. Isabelle Anguelovski et al., "Grabbed Urban Landscapes: Socio-Spatial Tensions in Green Infrastructure Planning in Medellín," *International Journal of Urban and Regional Research* 43, no. 1 (2019): 133–156.
10. Carlos Velásquez Mesa and Inter-barrial de Desconectados de Medellín, "Medellín Desconectada . . . De La Dignidad," *Observatorio K* 2, no. 1 (2010): 103–109.
11. "Siguen las versiones encontradas sobre fallas en la Biblioteca España," *El Tiempo* (Medellín), April 25, 2017, https://www.eltiempo.com/colombia/medellin/distintas-versiones-sobre-fallas-en-parque-biblioteca-espana-81368.
12. Andrés Sánchez Jabba, "The Reinvention of Medellín," *Lecturas de economía* 78, no. January–June (2013): 185–227.
13. Edgar Varela Barrios, "Estrategias de expansión y modos de gestión en empresas públicas de Medellín," *Estudios políticos* 36, no. S.l. (2010): 141–165.
14. Julio Dávila and Peter Brand, "La gobernanza del transporte público urbano: Indagaciones alrededor de los metrocables de Medellín," *Bitacora 21 Urbano-Territorial* 2 (2012): 85–96, 95.
15. Juan Carlos López Díez, "La gestión de la empresa pública: Lecciones de una empresa de servicios públicos," *Ad-Mniniter EAFIT* 7 (July–December 2006): 70–80.
16. See "Así se convirtió EPM en una multilatina pública," *Dinero*, August 18, 2016, https://www.dinero.com/edicion-impresa/caratula/articulo/como-se-convirtio-epm-en-una-multilatina-publica/228965.
17. See Caroline Merin, Alex Nikolov, and Andrea Vidler, "Local Government Handbook: How to Create an Innovative City," *Knowledge @ Wharton*, Wharton University

of Pennsylvania, December 20, 2013, https://knowledge.wharton.upenn.edu/article/local-government-handbook-create-innovative-city/.

18. Dávila and Brand, "La gobernanza del transporte público urbano."
19. See Merin, Nikolov, and Vidler, "Local Government Handbook."
20. See Juan Camilo Quintero, "EPM, Seis décadas de historia," *El Colombiano*, February 28, 2019, https://www.elcolombiano.com/opinion/columnistas/epm-seis-decadas-de-historia-DK10293312.
21. Luis Fernando Vargas Alzate, "Análisis de la política pública de cooperación internacional de Medellín," *Analecta política* 4, no. 6 (2019): 141–162.
22. "InterntionalCooperation," ACI Medellín, https://www.acimedellin.org/international-cooperation/?lang=en.
23. Julio D. Dávila and Diana Daste, "Pobreza, participación y Metrocable. Estudio del caso de Medellín," *Boletín CF+S*, no. 54 (2011): 121–131.
24. See "Historia," Metro de Medellín, https://www.metrodemedellin.gov.co/qui%C3%A9nessomos/historia.
25. See "La revolución de Fajarado," *Semana*, July 25, 2004, https://www.semana.com/nacion/articulo/la-revolucion-fajardo/67159-3.
26. Vargas Alzate, "Análisis de la política pública de cooperación internacional de Medellín"; "International Cooperation," ACI Medellín, https://www.acimedellin.org/international-cooperation/?lang=en.
27. An exception might be the Green Belt initiative that has displaced hundred of low-income families.
28. Gonzalo Lizarralde, "Decentralizing (Re)Construction: Agriculture Cooperatives as a Vehicle for Reconstruction in Colombia," in *Building Back Better: Delivering People-Centered Housing Reconstruction at Scale*, ed. Michal Lyons and Theo Schilderman (London: Practical Action, 2010), 191–214.
29. Lizarralde, "Decentralizing (Re)Construction."
30. Gonzalo Lizarralde, "Organisational System and Performance of Post-Disaster Reconstruction Projects" (PhD thesis, Université de Montréal, 2004), 460.
31. Arturo Escobar, "El desarrollo sostenible: Diálogo de discursos," *Ecología política*, no. 9 (1995): 7–25; Shiv Visvanathan, "Mrs. Brundtland's Disenchanted Cosmos," *Alternatives* 16, no. 3 (1991): 377–384; Gustavo Blanco-Wells and María Griselda Günther, "De crisis, ecologías y transiciones: Reflexiones sobre teoría social Latinoamericana frente al cambio ambiental global," *Revista Colombiana de Sociología* 42, no. 1 (2019): 19–40.

8. Humility: "The Damn Circumstance of Water Everywhere"

1. Unité de construction de logements et de bâtiments publics (UCLBP), *Rencontre de présentation du plan d'action pour la restructuration urbaine de la zone de Canaan et ses environs*, September 9, 2015, https://www.facebook.com/notes/uclbp/rencontre-de-pr%C3%A9sentation-du-plan-daction-pour-la-restructuration-urbaine-de-la-/732019023591228/.

2. Anne-Marie Petter, Danielle Labbé, Gonzalo Lizarralde, and Jean Goulet, "City Profile: Canaan, Haiti—a New Post Disaster City," *Cities* 104 (September 2020), article 102805.
3. I have intentionally modified the name of the city to protect the privacy of the local leader in this story.
4. We constantly design and regulate our own environments. Ever since the French philosopher and sociologist Henri Lefebvre published *The Production of Space* in 1974, scholars have been fascinated by how the city is created by ordinary citizens. They have studied, for instance, "architecture without architects" and "planning without planners."
5. Bernard Rudofsky, *Architecture Without Architects: A Short Introduction to Non-Pedigreed Architecture* (Garden City, NY: Doubleday, 1964).
6. In many developing countries, the informal economy accounts for more than 20 percent of gross domestic product.
7. For a detailed analysis of informal urbanism, please refer to Gonzalo Lizarralde, *The Invisible Houses: Rethinking and Designing Low-Cost Housing in Developing Countries* (London: Routledge, 2014).
8. In a study I conducted with a team from Université de Québec à Montréal (UQAM) in Port-au-Prince between 2015 and 2018, we found that one of the most common narratives of informality is its chaotic nature. In Haiti, informality is often associated with anarchy and chaos, and thus it is considered a problem that needs to be eradicated.
9. Here Lefebvre's work is particularly useful. It is closely linked to the notion of marginalized citizens claiming space in the city. Lefebvre sees space appropriation as a response to exclusion and domination—citizens claim their right to the city in a constant struggle against oppressive urban elites.
10. Georgia Cardosi and Gonzalo Lizarralde, *Understanding Urban Form and Space Production in Informal Settlements: The Toi Market in Nairobi, Kenya*, in *Proceedings of the 25th World Congress of Architecture: Architecture Otherwhere* (Durban, South Africa: UIA, 2014)
11. Stewart Brand, *How Buildings Learn: What Happens After They're Built* (London: Penguin, 1995).
12. See, for instance, the concept of "thirdspace": Edward W. Soja, *Thirdspace: Journeys to Los Angeles and Other Real-and-Imagined Places* (Cambridge, MA: Blackwell, 1996).
13. Thanks to Lefebvre and other urban experts, we know that urban space is a physical and mental creation. According to Lefebvre, social space is simultaneously perceived, conceived, and lived by humans. Thus, space is not only a material construct but also a social one. The concept of thirdspace, as explored by Edward Soja, a professor of urban planning at the University of California, Los Angeles, captures the physical presence of space and its simultaneous social and human value. According to this approach, when individuals design space, they express ideas and aspirations, provide meanings, apprehend symbols, construct narratives, and shape collective bonds.

14. This argument has been raised in several publications, such as Peter Kellett and Jaime Hernández-García, eds., *Researching the Contemporary City: Identity, Environment and Social Inclusion in Developing Urban Areas* (Bogotá: Editorial Pontificia Universidad Javeriana, 2013). The concept of aspirational design has been developed in Peter Kellett, "Original Copies? Imitative Design Practices in Informal Settlements," *International Journal of Architectural Research* 7, no. 1 (2013): 151–61.
15. Lizarralde, *The Invisible Houses.*
16. This argument is beautifully explained in: David Owen, *The Conundrum: How Scientific Innovation, Increased Efficiency, and Good Intentions Can Make Our Energy and Climate Problems Worse* (New York: Riverhead, 2012).
17. Stephen Pinker, *Enlightenment Now: The Case for Reason, Science, Humanism, and Progress* (New York: Viking, 2018).
18. Isabelle Anguelovski, "Environmental Justice," in *Degrowth: A Vocabulary for a New Era*, ed. Giacomo D'Alisa et al. (New York: Routledge, 2014), 65–69.

Index